Artificial Intelligence needs explanation

Silvie Spreeuwenberg

Colophon

Copyright 2019 by Silvie Spreeuwenberg
ISBN 978-90-815568-4-2
Graphic design by Kumi Hiroi
Illustrations by freepik.com
Kumi Hiroi, Anke Nobel (airport)
Picture by Ben Kleyn (cover)

Publisher: LibRT BV, Amsterdam
First print 2019

Visit www.ExplainableAI.info for
a list of related services such as training,
workshops and lectures.

As the years went by, I became convinced
that the influence of automatic computers
in their capacity of tools would only be
a ripple on the surface of our society,
compared with the deep influence they
were bound to have on our culture in their
capacity of intellectual challenge to Man-
kind that was totally without precedent.

Prof. Dr. Edsger W. Dijkstra,
Burroughs Research Fellow,
1975 EWD 533-0, The Netherlands

For my family, loved ones and friends who inspire me to be passionate without time and location boundaries.

Contents

Preface & acknowledgement

Economists say that we need to keep the money moving for the economy to flourish. I strongly believe the same is true for knowledge. We need to continue to share knowledge for societies to flourish. My motivation to write this book is to share my knowledge and encourage knowledge sharing within and between organizations.

The recent popularity of systems that claim to be artificially intelligent by having 'machine learned' knowledge gives the impression that even managing knowledge is a process that can be outsourced and automated. Every day I encounter opinions that say outsourcing knowledge creation to machines is scary. How do we control these machines? How dependent do we want to become and what does it mean for individuals, businesses and societies? Why is Henry Kissinger writing about a revolution driven by artificial intelligence? Why do the US, Europe and China compete in AI? What are the reasons that even the tech industry warns us about intelligent machines?

All my life I have been working on improving knowledge management processes. I thought it was time to share my ideas on these topics with a wider audience.

You may be someone with a general interest in Artificial Intelligence – AI – or be a business manager. You see the potential of AI systems for your organization and want to stay in control. Mainstream businesses are researching the potential of AI and want to do that in a controlled way. This book will provide you with a strategy to achieve trustworthy AI systems. Prior knowledge about AI or system development is not needed to benefit from this book.

I studied AI and graduated with a thesis on generating a human readable explanation for a machine learned model. To my surprise, generating explanations is still not a standard feature of the currently popular AI platforms. Since most businesses had no big data when I graduated in 1996, I focused on decision support systems based on rules. As a consultant, I worked mostly for large corporations. Typically, I operate in complex, knowledge intensive, environments: harbours, airports, insurance companies and tax administrators. I developed software, courses and methods to really engage business experts in making IT solutions. Many of the examples and inspiration to write this book are based on these experiences and the many interactions with businesspeople and other professionals.

I am grateful to many people who contributed to this book by discussing ideas, sharing experiences or reviewing the manuscript:

"There must be another way to do that."

ALCEDO COENEN
out-of-the box thinker and co-author of my first book on business rules, for placing constructive comments on my LinkedIn posts

ARJAN VREEKEN
lecturer Information Science at the University of Amsterdam, for contributing general examples and helping shape the book to enable its use as learning material.

"What does intelligence mean?"

DAGMAR MONET
Prof. Dr. Computer Science, HWR Berlin, for her effort defining intelligence, for sharing her experiences as an early AI adopter

EELCO DEN HEIJER
who studied with me and is an experienced software developer with a Master in Artificial Intelligence and PhD in using machine learning to generate visual art, for giving detailed feedback

EVERT HAASDIJK

a colleague while I worked on my thesis research on Genetic Algorithms, now Senior Manager and AI expert at Deloitte Nederland, for sharing his experiences

JAMES TAYLOR

leading authority on decision management, decision modeling, business rules and predictive analytics, author of the book Decision Management Systems, for sharing his ideas on decision analysis and allowing him to use his example

FRANS RAZOUX SCHULTZ

board member of the Dutch Association for Business Rules (BRPN) which I founded in 2005, ambitious and result driven Sales Manager specialized in consultative selling and new business, for discussing the initial concept

JAN VELDSINK

board member of the Dutch Association for Business Rules (BRPN), creative thinker, core teacher Nyenrode executive MBA for contributing examples to prevent cyber crime in financial institutions, his inspiration, references and examples of practical issues with AI for fraud detection

LINDA TERLOUW

a computer scientist and a professional independent Dutch business woman, for sharing her experiences with AI and image recognition to protect a society from harm

PINAKI GHOSH

smart listener who works on an intelligent house, inspired by my work on agent-based reasoning systems such as the rule-based traffic management, for his detailed comments

NATE MCCRAY

fellow student at the Erasmus OneMBA program, for discussing the promotion strategy

PIETER DEN HAMER

fellow student at cognitive AI in Utrecht, and now Senior Director at Gartner, for supporting the publication

ROB VAN HAARST

who shares my interest in controlled natural languages, and author of the book 'SBVR Made Easy', for contributing examples of how rules and AI methods are complementary

RONALD G. ROSS

visionary and partner in my first software company RuleArts, chair of the Building Business Capability (BBC) conference, recognized internationally as the 'father of business rules' and author of many professional books, for pre-publishing my ideas in the BRJournal and giving valuable feedback

SIJMEN RUWHOF

an ethical hacker, for discussing ethics and privacy protection laws

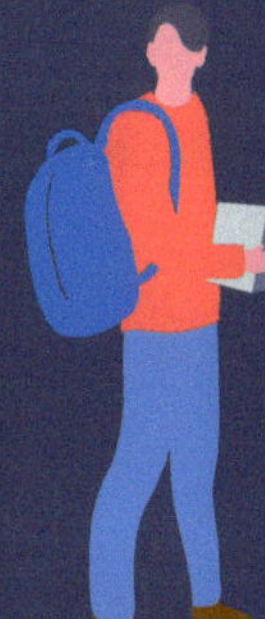

STIJN HOPPENBROUWERS

professor (Data & Knowledge Engineering) at Hogeschool van Arnhem en Nijmegen (HAN), for providing feedback

ROSALIE VAN OOSTROM

responsible for defining and implementing trustworthy AI at the Dutch Ministry of Health, experienced entrepreneur and a legal expert with beta brains, for giving details about regulations that already require explainable decisions

TOM VAN ENGERS
fellow student at cognitive AI in Utrecht,
professor at University of Amsterdam
and director of the Leibniz institute of law,
for inspiring me with his publication on
normware

Finally my special thanks to:

PATRICIA HENAO
program manager for one of my first large
knowledge management projects, a professional independent instructor, certified
Project Manager and Facilitator, for her
critical reading

LAURA GERRITS
my daughter, for being patient when I had
to 'finish another sentence'

1

Choose for AI and explainability

Artificial Intelligence (AI) is increasingly popular in business initiatives. It has the potential to disrupt the healthcare and financial services industries. Corporate business functions, such as finance and sales, have already changed. Did you know that startups in AI raise more money than other startups? And that, within the next decade, every individual is expected to interact with AI-based technology on a daily basis?

AI is a technology trend covering any development to automate or optimize tasks that have traditionally required human intelligence. Experts in the industry prefer to nuance this broad definition of AI by distinguishing machine learning, statistics, IT and rule-based systems. In this book I will use the term AI to reference all techniques popular today under this name. The availability of huge amounts of data and more processing power – not major technological innovations – make machine learning the most popular technique today. However, I will argue in the remainder of this book that the other AI techniques are equally important. While the popularity of AI has triggered me to write this book, most conclusions are applicable to any IT system.

Since the invention of the computer, AI and IT have been closely related. Think of Lady Lovelace, who worked in 1837, together with Charles Babbage, on the first design of a programmable computer. Back then it was known as an 'algebraic machine' and she named it The Analytical Engine.

Lady Lovelace was the daughter of the romantic poet Lord Byron, and was a gifted mathematician and intellectual. When she translated an Italian article on algebraic machines, she supplemented it with extensive notes on such machine's capabilities. These notes, published in 1843, prove that she was the first to record that a machine could be programmed to solve problems of any complexity or even compose music. This formed the inspiration for theories on logic that resulted in the programming

ADA LOVELACE

A PICTURE OF THE ANALYTICAL MACHINE DISPLAYED IN LONDON.

The description generated by Microsoft's AI engine is as follows "A picture containing wall, indoor, chair, table". This is not the right description and illustrates the limitations of AI generated image descriptions.

languages in use today.

Lovelace died early and the Analytical Engine was not built during her life time. Her soulmate Mr. Babbage did create a trial model of the Analytical Engine that is displayed in the Science Museum in London.

She is regarded as the first computer programmer but was most likely thinking about AI, or what we would call AI today, when she said: "The Analytical Engine has no pretensions whatsoever to originate anything. It can do whatever we know how to order it to perform. It can follow an analysis, but it has no power of anticipating any analytical relations or truth."

Her statement has been debated by Alan Turing, another example of a person in which the close relationship between AI and IT comes together. He defined a test, named after himself, of a machine's ability to exhibit intelligent behavior. The test is an experimental setup. Someone judges a conversation without knowing whether the conversation is with a human or a machine. All participants are separated from one another. The Turing test is successful if none of the participants is able to tell the difference between communicating with the machine and communicating with the (other) human. This test, in many variations, still plays a role in defining AI.

At the same time Turing played a crucial role by creating one of the first computers based on the Von Neuman design: a computer that had a stored-program. This marked the beginning of programmable machines, the start of executing the vision of Lady Lovelace and the rise of the software industry.

The consequences of this innovation for humanity have been huge and were, at the time, difficult to oversee. There were pioneers, visionaries, investments and failures needed to get us to where we are today. I am so grateful with the result. Every day I use a smart phone to provide me travel advice, ways to socialize and recommendations on what to buy. Computers also help me memorize and acquire new knowledge. Many of these innovations are related to technology developed by researchers in Artificial Intelligence. The full potential has not yet been exploited.

But there are also concerns.

Artificial intelligence solutions are accepted to be a black box: they provide answers without a motivation, like an oracle.

You may already have seen the results in our society: AI is said to be biased, as a result, governments raise concerns about the ethical consequences of AI and regulators require more transparency. We should embrace the potential improvements that AI could bring to improve human decision-making. Instead people have become skeptical about AI technology. You may fear losing your job. Or, as a specialist, you are aware of all the uncertainties that surround your work. How can an AI algorithm deal with those aspects?

Perhaps you guide your company to compete using AI. What approach could you follow without losing the trust of your employees or customers?

Now that AI technology is at the peak in the hype cycle for emerging technologies more conservative businesses want to use the benefits of AI based solutions in their operations. However, they require an answer to some or all of these – above mentioned – concerns.

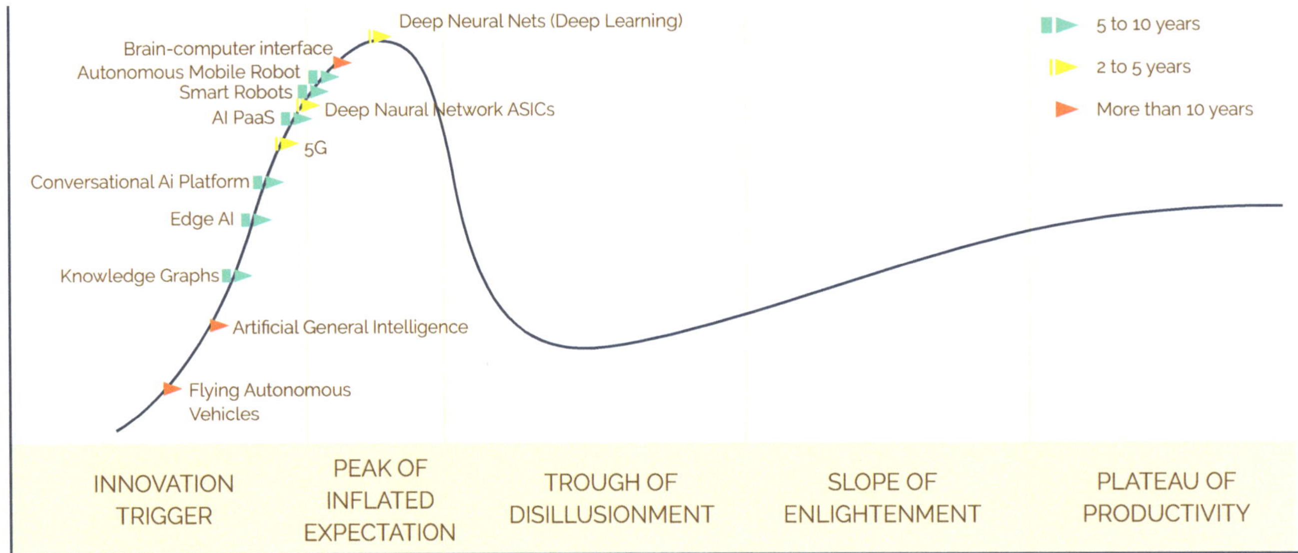

To benefit from the potential of AI the resulting decisions must be explainable. For me, this is a no-brainer since I have been promoting transparency in decision making for years, using rule-based technology for years. In my vision view, a decision support system needs to be integrated in the value cycle of an organization. Business stakeholders should really feel responsible for the knowledge and behavior of the system and be confident of its outcome. This may sound logical and easy, but everyone with experience in the corporate world knows it is not. The gap between business and IT is filled with misunderstandings, differences in presentation and expectations.

It takes two to tango.

Business, represented by subject matter experts, policy makers, managers, and executives, should take responsibility for their knowledge. IT should create integrated systems that support decision-making based on business knowledge. Explaining these decisions plays a crucial role in such collaboration.

We should do the same for AI based decisions: choose AI technology when needed and use explanations to make it a success. That is, explainable AI – known by the acronym 'XAI'.

The five reasons why XAI solutions are more successful than an oracle based on AI, or any black box IT system, are as follows:

1. Decision support systems that explain themselves have a higher return on investment because explanations close the feedback loop between strategy and operations resulting in timely adaption to changes, longer system life time and better integration with business values.

2. Offering explanations enhances stakeholder trust because the decisions are credible for the customer and also makes a business accountable towards regulators

3. Decisions with explanations become better decisions because the explanations show (unwanted) biases and help to include missing, common sense, knowledge.

4. It is feasible to implement AI solutions that generate explanations without a huge drop in performance with the six-step method that I

developed and technology expected from increased research activity.

5. To be prepared for the increased demand for transparency based on concerns about the ethics of AI and the effect on the fundamental principles of a democratic society.

Each chapter will explain one of these reasons in detail and provide examples or practical guidance. After reading you will have a good understanding what it takes to explain solutions that support or automate a decision task and the value explanations add to your organization.

I have tried to stay away from technical details. When I describe how to implement XAI solutions, in chapter 5, I make an exception to this self-imposed rule and introduce some details about the underlying technologies. Just enough for a good understanding, not enough to 'do it yourself'. In the final chapter I recommend sources to use if you need help understanding the underlined topics in more detail.

2

Systems that explain themselves have a higher ROI

2.1 Explanations make a solution easier adaptable to changes

2.2 Explanations increase the success rate of AI solutions

Most technology used today for AI already existed when I graduated in Artificial Intelligence. However, unlike today, there was no big data to apply machine learning methods and computer processing power was about 20% of the current state. After graduating, my job was to create 'expert systems': systems that make a decision based on a set of rules. To apply rules, you need structured data. Most of the time there was no structured data. I even had to introduce the first database to my client organizations.

Expert systems where part of the second wave of AI hypes that made similar claims as other later hypes in the software industry: higher efficiency, better performance and easier maintenance.

My first job was to train subject matter experts (SME's) to write rules. It was a disaster. The rules were incomplete, inconsistent and if they finally had a set of rules that produced the right result, they were redundant and impossible to manage. I had to conclude that it was difficult to get the rules right and experts did not see it as part of their job to be responsible for the rules and behaviour of an IT system. The troubled relationship between the business and IT were both the cause and the result.

I did not give up. Since that first experience my objective was to develop methods that make it easier for the business to understand, manage and take responsibility over system's decisions. I use decision tables, decision trees, controlled natural language, domain specific languages and any other form of good meta modelling. Which one is best depends on the problem characteristic, the domain, and of course, what the subject matter expert is able and willing to do.

Rule based decisions are relatively easy to explain. However, I have seen many organizations that did not use this benefit to its full potential. Instead they acted slowly to adapt out-dated rules, while operations developed 'workarounds' and labelled the decision support system as 'legacy'. This results in a bad return on their investment - ROI. This nightmare is easy to repeat with new AI technology because there is no fundamental difference between AI and software development.

Each AI solution is also an IT solution and there have been many issues reported on IT projects that fail after millions have been spent or do not return the expected values. The general issue with the success rate of IT is that people overestimate their ability to evolve the systems simultaneously with the business environment. At the same time they overestimate their IT skills, but that is quite another discussion. AI will not change this. The bottom line is that we can't just use AI instead of understanding but must understand first and then add AI.

In the famous article "No Silver Bullet" by Brooks in 1986, he argued that "we cannot expect ever to see two-fold gains every two years in software development, as there is in hardware development". AI and XAI will not change this because his arguments still hold today.

Brooks claims that there is a fundamental complexity in requirements engineering for software. Specifically, the right level of detail, interaction with the user and explanation model differs per industry, topic, technology and decision characteristics, therefore, good engineering practices still apply to AI. "Techniques for expert systems, image recognition and text processing have little in common", he states, "therefore, the work [in AI] is problem-specific and the appropriate creative and technical skills are required to transfer the [AI] technology into a good application."

His recommendation to "grow" software organically, through incremental development is compatible with the agile methodologies that are popular today and the feedback cycles suggested by the guidelines for trustworthy AI. Explanations play a crucial role in this feedback cycle by indicating when a new cycle is needed to adapt the system to changes.

2.1 Explanations make a solution easier adaptable to changes

Explanations inform us about the distinguishing aspects a decision is based upon. Explanations also indicate when a certain case is an 'outlier' and very different from the data used to train a model. Hence explanations help us to understand when it is time to adapt.

A plan-do-check-act cycle (PDCA) is the business variant of a feedback loop in an engineering setting.

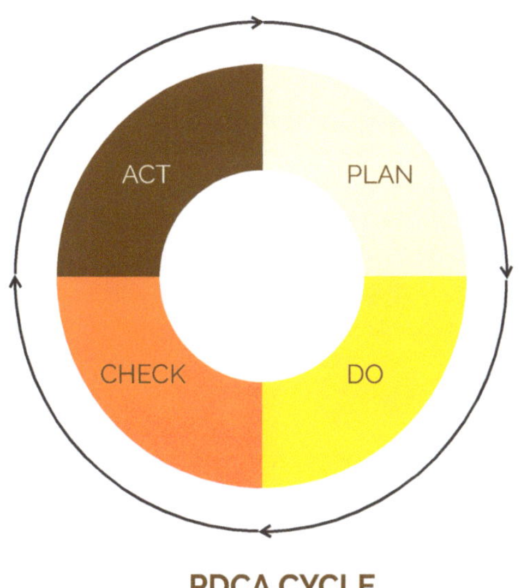

PDCA CYCLE

In my experience, the business variant in many organizations is an exercise on paper instead of a real-world practice. What is missing? When compared to the engineering setting, the PDCA misses two elements: a clear-cut answer telling us when outcomes are 'an error', and a mechanism, or controller, to correct the system based on the error. What is missing is good feedback.

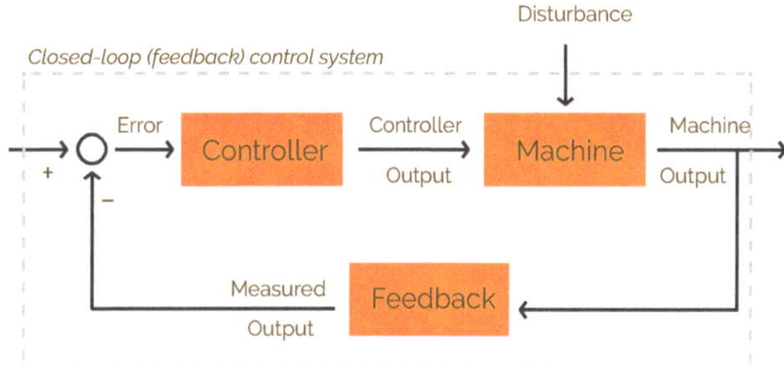

ENGINEERING FEEDBACK LOOP

We need systems, or ecosystems, in which both elements are clearly present. That is not to say that every step has to be automated. It is perfectly fine when there is 'a human' in the loop.

Most important is that the feedback mechanism is monitoring all the norms on operational, tactical and strategic levels. The system monitoring these norms referred to as <u>normware</u>, has a well-defined role: the ability to measure in an ecosystem supporting the full decision-making process. It is an interface between the operational level, the directors or policy makers, the managers and, potentially, an AI-solution.

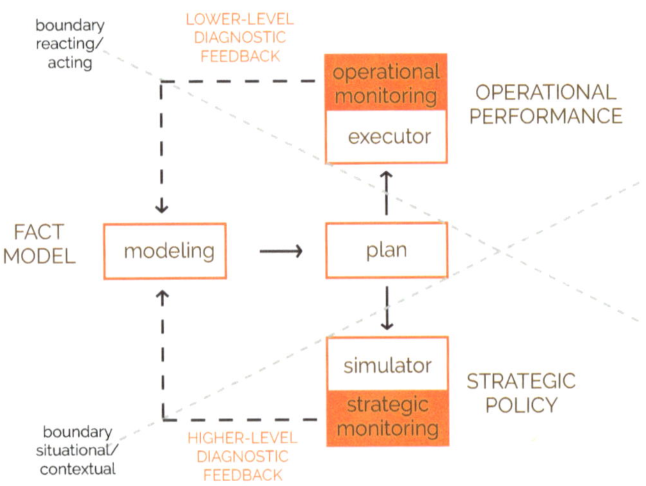

NORMWARE

A full decision-making process supports learning new knowledge. We learn when the environment or our strategy changes and we adjust our decision-making norms accordingly. The process is the same when we use data to train an AI model that is used to make a decision.

Timely modification of systems prevents users from creating 'workarounds', keeps the system up-to-date and extends the life of the system, resulting in a higher ROI.

2.2 Explanations increase the success rate of AI solutions

IT and AI are closely related because both are used to automate a task, using a computer, which would otherwise have been carried out by a human being or would require an infinite number of personnel. While most IT systems are not AI, AI systems are IT systems and AI research is not IT. Do you follow?

MODERN RESOLUTION FOR IT PROJECTS

	2011	2012	2013	2014	2015
Successful	29%	27%	31%	28%	29%
Challenged	49%	58%	50%	55%	52%
Failed	22%	17%	19%	17%	19%

The numbers in the above graph, based on the CHAOS Report from the Standish Group, list the percentage of IT projects from 2011 – 2015 that are delivered On Time, On Budget, with satisfactory result (called the Modern Resolution).

As many organizations already struggle to integrate IT in their daily strategy, and AI is IT, AI investments have a high-risk profile. Additionally, companies face major challenges to get a good return on IT projects and 20% of IT projects fail according to the numbers of the Standish Group.

One of the main causes is a broken relationship between the IT and operations departments. Without basic digitization skills management cannot lead and must rely on IT suppliers. The organization cannot learn because IT emphasizes the complexity of system development. IT suppliers benefit from opaque systems and maximize profits with spending hours. At the same time, the business is happy to hide its mistakes by handing over its responsibility to IT. The result is a troubled relationship between the business and IT that hinders a creative and curious mindset.

In such a troubled relationship nobody benefits from answering the fundamental question: why does the system behave the way it does? If we don't know how the system operates how can we expect IT and business stakeholders to collaborate, work iteratively and create real value for the business? By requiring explainable AI, managers can change this situation and improve their return on IT investments using AI.

How would that work in practice? Over enthusiastic techies promote AI that is able to perform any intellectual task that humans can do, even in a changing environment. Such AI is called 'strong AI'. After an initial substantial engineering effort, it would require little maintenance.

Who answers the question: "Why does the system behave the way it does?"

The AI community distinguishes strong AI from 'narrow (or weak) AI', which focuses on automating just one task. So far, this is what we see all around us and it requires a good understanding of a well-defined task with some initial engineering effort.

meaningful way when the task at hand does not fit in the scope of the AI solution. In this model there is a continuous engineering effort. Business stakeholders use explanations to understand the system and feel responsible for the knowledge and behavior of the system. The AI solution is integrated in the essential value creating cycle of an organization. The relationship between IT and the business is improved and this leads to less high-risk profile IT investments.

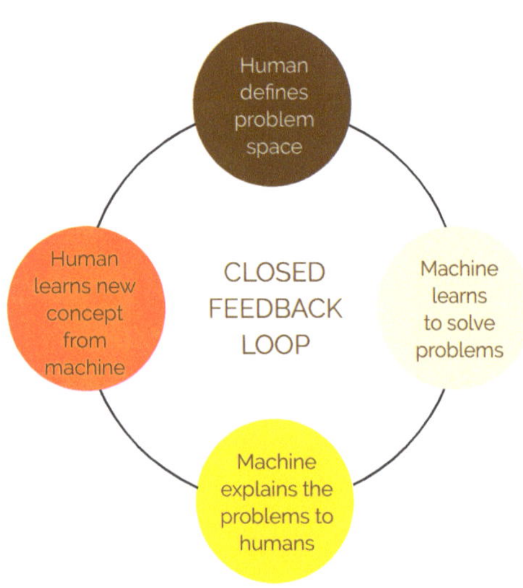

XAI aims at something in between narrow and strong AI. It is still narrow AI but used in such a way that there is a feedback loop to the environment and the feedback loop may involve human intervention. We understand the scope of the narrow AI solution. We can adjust the solution when the task at hand requires more knowledge. We are warned in a

3

Your business will be trusted

If you start thinking about the people you trust in your environment, it's likely that you end up more confused than enlightened. That is what happened to me! There are people whom I have trusted right from the beginning, people who have gained my trust over time, and people whom I trust but have habits that are a bit annoying, like always being late. Although I know perfectly when I trust someone and when I don't, it's a difficult concept to define.

The idea of 'trust' includes a couple of concepts that interact with each other. One way to describe this interaction is using the following trust-equation as defined by Charles H. Green:

trust = (reliability + credibility + intimacy) / self-orientation

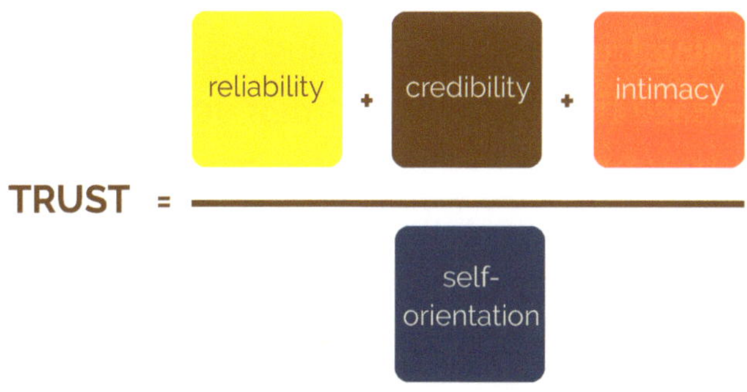

The equation helps me to understand the differences between the people I trust. One of them may not be very reliable – being late - but makes up by having a lot of credibility – being very smart and knowledgeable. When the person is not only connected to me, but also to other people I am close to, therefore more intimate, this increases my trust in this person. However, if the person has all these characteristics but acts out of self-interest, is self-oriented or does not really listen to me, my trust in the person decreases.

Trust in an organization is an important factor for business continuity. Consider the loss of trust related to recent events involving Facebook, Volkswagen and Boeing. Similar to trusting people, trusting organizations is related to their credibility, reliability, and orientation.

Explanations play a crucial role in gaining trust because it helps organizations increase the credibility of their decisions. Furthermore, when decisions can be consistently explained organizations show that they are responsible and accountable. This increases our perception of their reliability and show an outward-orientation- two factors that increase trust.

3.1   Making credible decisions

No matter how 'intelligent' your machine is, a person or organization is still accountable for any decision that is supported by AI models. Therefore, these decisions should be acceptable and convincing for them. This means that the explanation of the decisive factors must meet the expectations of the user and make sense. If they do, we are happy to accept the credibility of the

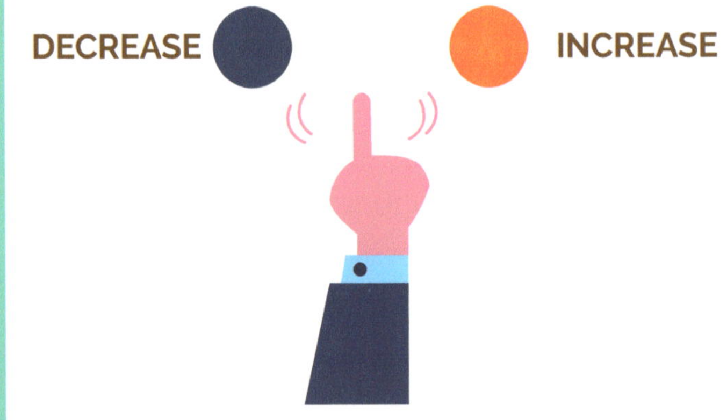

machine-learned decision and the confidence between the trustee (employees or customers) and trusted (decision making software) increases.

The other way around, when the outcome of a decision does not make sense or is not based on the right features – as is the case in the example where the passenger forecast at an airport does not take holidays into account – people will look at the decision as 'advice' and make their own decisions. The end result is that the machine-made decision is disregarded.

3.2   Being a responsible business

Making credible decisions is highly related to the trust that employees and customers have in your business.

Businesses also have to demonstrate compliance to external regulators. For example, regulators will mandate that 'life changing' decisions are explainable. Being able to consistently show a full trace, from each decision to policy and regulations, is already a necessity to be compliant in many industries like finance, pharma and health care. More regulations are expected due to the ethical concerns that I outline in chapter 6.

> **"The financial sector is commonly held to a higher societal standard than many other industries, as trust in financial institutions is considered essential for an effective financial system."**
>
> **– The Dutch Financial Authority (DNB), report on the use of Artificial Intelligence in the Financial Sector, 2019.**

Organizations can satisfy a regulator in different ways. I see many organizations creating compliance dashboards to report about the compliance of past transactions. They prove compliance after the fact. There are also organizations that use only technology that is certified. They prove compliance, in advance - before the fact. I believe organizations should include better explanations for *every* decision by involving business stakeholders and making them responsible. This way, you prove compliance during the decision-making process and the closed feedback loop, proposed in chapter 2, runs faster.

A financial institute uses <u>machine learning</u> methods to find patterns in transactions that are abnormal and may be fraud, or may indicate potential illegal money laundry.

The technique used is also named 'anomaly detection'. Transactions may be flagged, and the involved accounts may be blocked.

Customers and the authority for financial markets (AFM) require a well-defined reason for a blocked account. The reason: "a deep <u>neural network</u> scored the transaction above our threshold" will not satisfy them. So, all machine learned models must be validated, transformed to rules and implemented. The rules provide a rationale for flagged transactions.

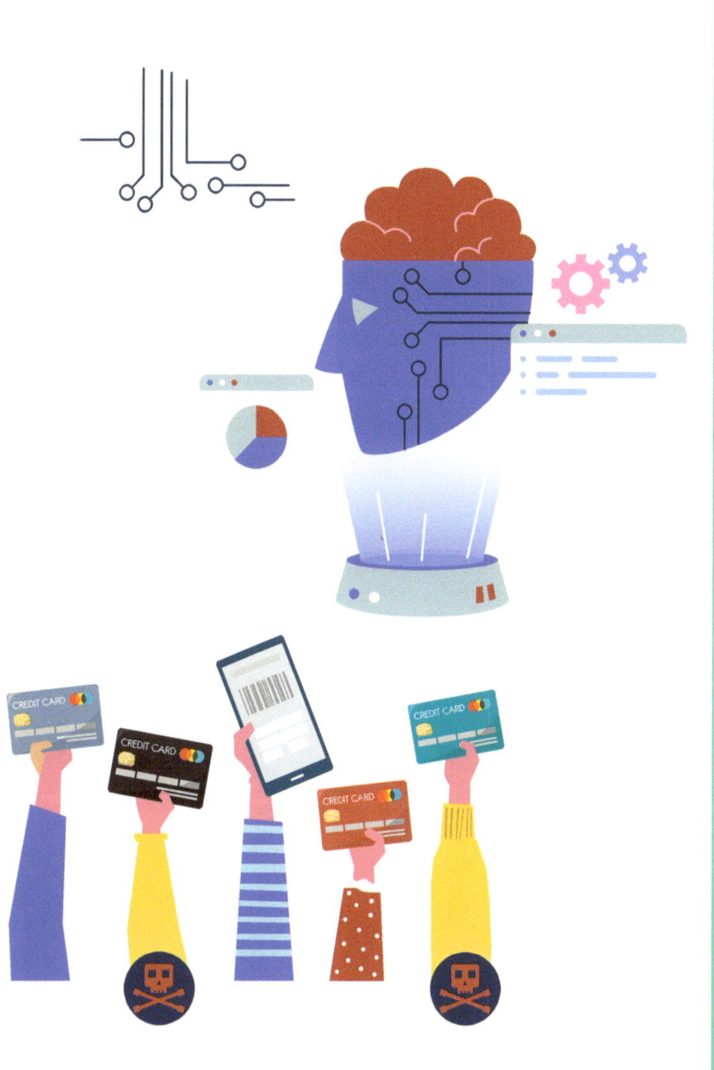

4

Explanations enhance decision making

4.1 Unwanted decision biases come to the surface

4.2 Missing commonsense knowledge is detected

The previous chapters demonstrate that I consider decision-making as a process that is embedded in a decision management process. An individual decision is taken at one point in time, however, the way in which we make decisions typically evolves over time. This evolution is the main objective of the decision management process and it is a continuous effort. An example description of such process is the OODA loop, developed by military strategist John Boyd, applicable to individuals, systems and organizations.

Research has demonstrated that human decision-makers have different kinds of biases and may also introduce those biases

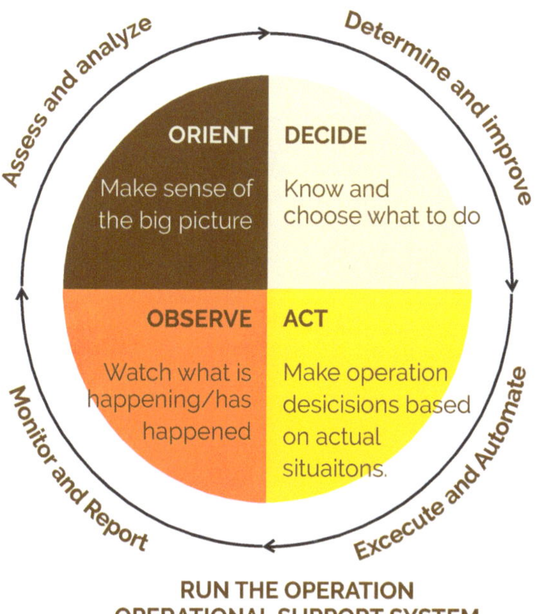

in system solutions. So, it is no big surprise that data collected in the past from human decisions is biased and that this bias is picked up by an AI algorithm resulting in machine-made decisions that include unwanted biases.

The famous example of a gender bias in HR matching software not only demonstrates a weakness in AI techniques, but also a weakness in humans. These biased decisions, whether human or machine-made, will need to be improved and explanations may help us to understand what a model learned, recognize the bias and improve the knowledge and decision-making of the model.

Unfortunately, the AI community has focused on optimizing the accuracy of the resulting models – which is important – and forgot to explain the results – which is not trivial. Recent developments in image interpretation confirm that we cannot always understand what the model has learned. In some cases, we may not even have the words to express the concepts that some parts of the model represent. Is that acceptable?

Human decision biases are well researched in economy and psychology. I investigated potential decision biases for gate planning at an airport. There are four major categories:

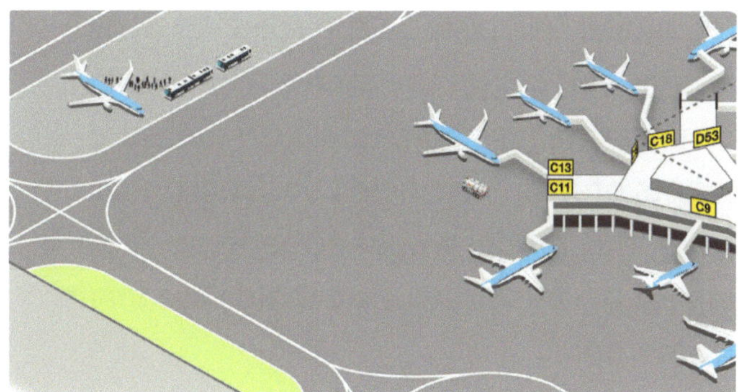

Salient information: People have a tendency to give more weight to prior experiences that stood out, like an incident or unexpected event, than is statistically justified. For example, if a similar flight from the same airline had a major delay the other week there is a tendency to plan the flight in such a way that a potential delay will not have a major impact on the overall planning. This tendency explains why there are differences between the plans of different gate planners.

Illusion of control: People have the tendency to believe they can control outcomes that they demonstrably have no influence over. Gate planners assume that by making a robust and practical planning they control a flight's departure. However, the reality is that they do not have control over many aspects that influence on time departure like the timely arrival of passengers, baggage, staff, etc.

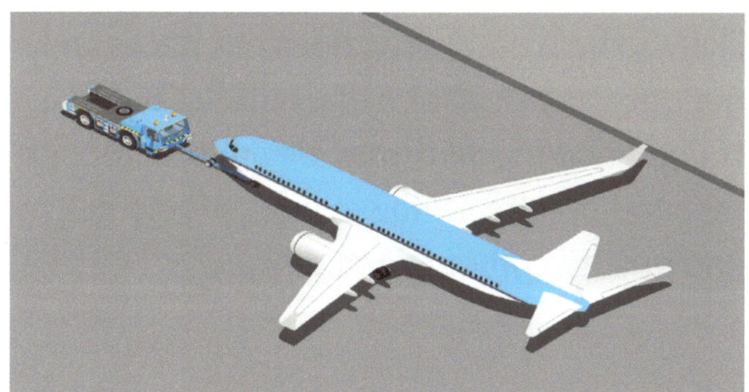

Conservatism: People generally fail to change their opinions when they receive new information. For example, there are some general opinions about the reliability and capacity of the handling agency. Therefore, a flight sometimes waits at a gate because ground control assumes that the handling agency has limited tow capacity and low tow speed.

Fundamental attribution error: There is a tendency to over-emphasize personality-based explanations for behavior while underemphasizing situational explanations. When faced with a decreasing performance, there is a tendency to look at factors that changed simultaneously (like the number of gate changes) or the personality of the gate planner – they like to puzzle the whole day – while the situational circumstance (more flights, bigger flights and new policies that affect the capacity) are overlooked.

Here are two cases that illustrate the importance of understanding what the model has learned.

Would you use an algorithm that recognizes dogs in pictures based on features in the environment, like trees, a tennis ball or a dog house, instead of aspects like 'is hairy', 'has a muzzle/snout', 'has large ears'?

A self-driving car uses a model trained to recognize "rubbish on the road" to avoid hitting objects on the street. The input is camera images combined with measurements of magnetic fields. In Japan, a car using this model would not enter any parking lot because their thresholds are made from iron.

There has been considerable research conducted that examined if the knowledge learned by a neural network reflects deep knowledge, abstraction and generalizations. The conclusion is that these non-human decision makers are easily fooled and likely to be biased. That does not necessarily need to be a problem: machines may be retrained and usability may be limited, though fast and practical. However, it should be known and communicated.

Searching on internet is text-based while much information on internet is visual. To find the right images and help visually impaired individuals to understand web pages, these images need to have a description. But we are lazy at making good descriptions for every image. Could we make an algorithm to generate them?

AI technology is used on a large scale to solve this problem. <u>Machine vision</u> tasks, based on value pairs – lots of images with matching descriptions – are reported to be 90+% accurate. That is good news for those who are visually impaired and have nothing to rely on now.

What did the model really learn? The answer to this question places the initial results in a different light. Typically, a system learns what humans captioned manually, and it will re-use it when presented with scenes similar to what it's seen before. When an image is presented that is not similar to anything presented before, the system will choose the caption that is most frequently used by humans (for example 'an animal in wild nature').

Despite the remarkable increase in the number of image description systems in recent years, experimental results suggest that system performance still falls short of human performance when presented with a foil.

A model that was trained to recognize porn in pictures classified all pictures with a woman as porn… this machine trained model learned the wrong thing.

A model that was trained to predict the departure time of a flight only learned that flights with more seats are likelier to be delayed even when these seats are not occupied by passengers …. this machine trained model learned a trivial thing.

If you are interested in this topic, refer to the recommendations and sources in chapter 8 on <u>machine vision</u> for readings that support this claim.

4.1　Unwanted decision biases come to the surface

Explanations will not directly prevent biases in AI models, however the explanation of a decision may reveal that the model picked up patterns in the data that are the result of unwanted human biases. Detecting these biases is important because as we have seen in recent examples, – that don't like to be named – were blindly trusted algorithms have the ability to damage your company's reputation.

Researchers have proposed solutions to remove or detect decision biases that have been demonstrated to work and should be included in your AI solution.

One solution is to avoid the 'solitary decision maker' and instead use our reasoning skills in a conversation to produce and refute arguments. Research proves that a decision-making process that confronts different biases - by bringing a diverse group of people together - limits the impact of the biases. Explanations play a role in such process by facilitating the conversations between the machine, the subject matter expert and the person or client that is affected by the decision. Another solution is to include a system of checks and balances at three levels: Operational, Tactical and Strategic.

-   At a strategic level, we should monitor our KPIs to decide when constraints, processes or guidance needs to be adjusted or "corrected."

-   At the tactical level, the strategy is transformed to a model with a desired outcome, that must be tuned, preferably based on parameters that influence your KPIs.

-   At the operational level, the error between the actual and the desired outcome must be checked, using diagnostic feedback and explanations.

The idea of <u>normware</u>, introduced in chapter 2, connects these three levels, which often act as silos in an organization. The introduction of a strategic decision-making process based on normware, will help people come together with a different angle of vision, make uncertainty explicit and depersonalize strategic decision making. These are all proven factors to reduce biases in strategic decision making.

4.2 Missing commonsense knowledge is detected

A decision model that fails to include commonsense knowledge will, in the end, be outperformed by humans. Examination of the explanation for a decision may quickly reveal that some knowledge is not taken into account and that the model must be updated. This is how explanations can help to identify common knowledge needed in decision support systems.

XAI solutions should have a mechanism to include such commonsense knowledge in decision support systems. Using more statistics is not effective, as one snowy day a year will not affect averages and correlations very much, and machine learned models fail to deal with exceptions.

Instead I expect – and you should ask for – hybrid solutions using both machine learning based on statistics to deal with uncertainty, and reasoning based on rules to deal with commonsense knowledge. Further, we expect a supporting mechanism based on good explanations, to help the business recognize that knowledge may be missing.

The next chapter will provide guidelines on how to create such eco-system as part of a six-steps method.

To accurately forecast the number of passengers on the first day of a holiday season you expect this common knowledge to be part of the explanation. If not, you may want to deviate from the forecast and may have a valid reason to do so.

The explanation must show the causality relationship between concepts. The cause of an increase in passengers may be the start of spring break in the south of the Netherlands. As a result, there will be a correlation between passengers from the south of the Netherlands travelling to southern European destinations and the increase in the number of passengers. Needless to say, neither the origin nor the destination are is causing the increase in passenger numbers. To detect such flaws in cConceptual modelling, as presented in chapter 5, is of paramount importance to detect such flaws.

5

Explainable AI is feasible using a six-step approach

If you believe the reports, AI is just upload-
ing big-data, running an AI algorithm and
getting results. Unfortunately, it is not that
simple. Practical methods to apply AI are
often limited to guidelines on choosing the
right AI algorithm. As a result, AI solutions
are black boxes, with unclear scope, that
are not trusted by the workforce nor integrated in the organization's strategy cycle.

This chapter provides practical guidelines
to create a sustainable AI solution that is
integrated in the organization's strategy
development cycle and uses explanations
to guide humans in a plan-do-check-act
process. The result is an XAI solution combining symbolic reasoning – <u>business rules</u>
- to deal with commonsense knowledge
and <u>machine learning</u> - statistics - to deal
with uncertainty. In AI terms this is named a
'hybrid solution'.

A nice example that illustrates the challenge is
the Titanic dataset. Take a look at Kaggle.com.
You'll find a simple dataset of visitors on the
Titanic. The objective is to predict who survives
and who does not. There are features like gender, class, age etc. however, the best predictor
turns out to be the surname! You would hope to
find knowledge that could help people in a similar situation survive.However, the best predictor
turns out to be the surname!
Obviously, that knowledge is not applicable to a
new situation. There is no smart software available that can draw a more nuanced conclusion:
for example, that people tend to help a family
member, hence traveling with more family mem-
bers increases your chances of survival.

The six steps I present are a generalized method and blindly following any method is a recipe for disaster. Understanding the underlying motivation and thinking how to best apply it in a specific situation is a better approach. Furthermore, the steps need not be applied linearly; they may be revisited as often as needed. Instead of stepsyou could call them activities or actions.

The 6 steps that should be in your work cycle help you all the time to:

1. Get a shared understanding of the domain
2. Understand the task and select the right scope
3. Collect the right data and improve its quality
4. Select AI techniques that deliver results
5. Generate good explanations
6. Evolve the solution over time

Maybe you expected steps like: write use cases and user stories, create a minimal viable product – MVP -, integrate the MVP in IT infrastructure, refine the business process and validate performance. This is the typical terminology used in the AI world today.It aligns well with the "fail early, learn fast" culture, typical for startups (The Start-up Way) and promoted by venture capitalists. While this way of working is not wrong, we do need a different focus to create an explainable AI solution. Use cases and MVP's focus on the delivery of a product. Of course, that is important too.

We don't want decision support systems that result in the headline: "AI model of YOUR COMPANY NAME uses last name of applicant to determine insurance eligibility, is that legal?"

However, I prefer to focus on integration into the broader context of sustainable business operations and strategy development.

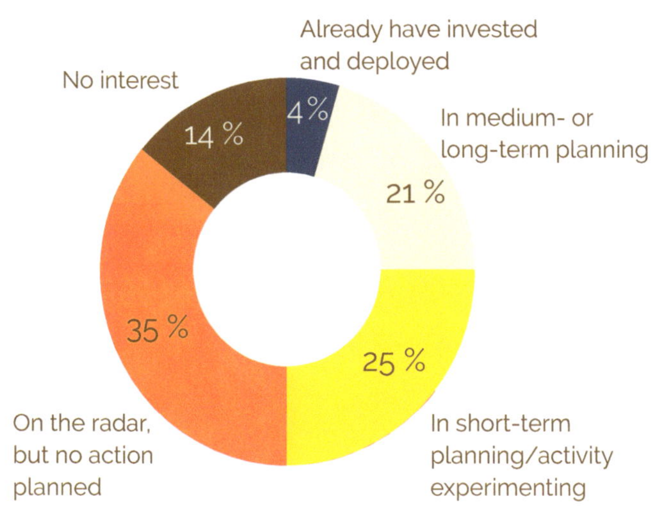

AI adoption has tripled in 2018 moving AI towards the Gartner-hype-cycle peak. Now that AI is getting more mainstream, 50% of business are planning AI initiatives. More conservative companies have good reasons to enter this arena and they will ask for methods that result in more reliable outcomes.

5.1 Get a shared understanding of the domain

It all starts by agreeing, as a business, on the meaning of the important concepts in your domain that trickle down from the organization's goals and objectives. Businesses that can state clearly who they target and what it means to be successful are likelier to be efficient, effective, and successful in the application of AI and decision support systems.

The important concepts of your business are the ones that change over time. An order becomes a delivery, a lead becomes a customer, a customer becomes a returning customer, a disease is cured, a loan application is approved, etc. Businesses exists because they are good in managing the change: they deliver on time, they understand customer needs, they know how to cure a disease, and they predict the return on a loan. AI technology is used to improve business decisions in this process: when can we deliver the order, what are the customer needs, what is the best cure, how risky is a loan? We use verbs to naturally relate these concepts with each other and with characteristics of the concepts. A customer has an age, the age of a customer influences the products that the customer may buy, etc.

In complex domains it may help to visualize the concepts, and their interrelationship, in a graphical model, for example using the fact-modelling approach. The result is not a data model! Instead it is a <u>concept model</u> showing the structure of meaning used by people to understand a domain.

The concepts in a concept model go beyond the data that is collected in a database and often involve abstractions of generalized concepts. For example, 'assessment' and 'genre' in the concept model of a Library example.

These concepts are as important as know- ing the meaning of the data that you collect. A clear understanding of relevant abstractions will help you in step 4 to select the right AI technology and create an AI architecture that makes sense.

Different departments, in an organization, often end-up with different models about the same domain. Such ambiguity can lead to misunderstandings and inefficient operations. We don't want to include ambiguity in our explanations! AI solutions often have the ambition to align a value

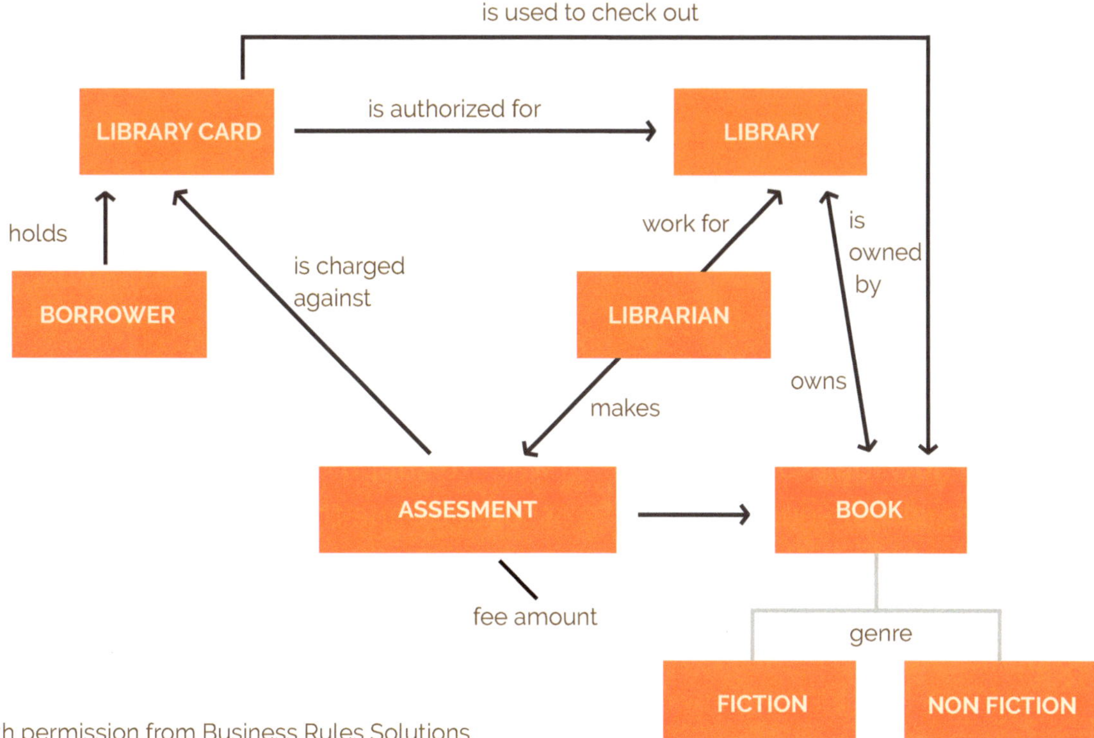

Used with permission from Business Rules Solutions

chain involving different departments, therefore it is a good idea to align the terminology between these departments as well.

In my experience, businesses that have well defined concepts are better in communicating with their customers, partners and technology departments and are more successful in realizing technology projects, including AI innovations.

Perhaps your organization is working on an enterprise ontology or data dictionary?

These are good initiatives to align terminology between departments. Hopefully your business stakeholders are involved!

5.2   Understand the task and select the right scope

We make decisions all day long and most of them are unconscious, like a habit. Think about the number of decisions that you make in the first hour after waking up: you decided to snooze the alarm; then you decided to not snooze again but to get out of bed; maybe you reminded yourself of an important meeting and about being on time; you decided on how long to shower, what to wear, the strength of your coffee, your mode of transportation to the office, the best time to leave to be on time ... it's amazing if you think about it.

Normally we don't think about all these details – they are left implicit -, but if we want to automate a task, we need to know them and this makes software engineering difficult. Engineering is based on rigorous methods applied to well-defined situations; when engineering is applied to decision making, we need to know the possible outcomes, the criteria used, and the context in which the decision is made.

Given that many decisions are made unconsciously, it might be a challenge to define a task sufficiently well. Luckily, we have thousands of years of engineering experience providing us with knowledge representation techniques, knowledge elicitation methods, gamification, facilitation and statistics to help us.

Following these practices you start by understanding the decision task that you want to support. This task defines your scope. Define the task using a question that represents the decision.

> **A good question starts using the right subject:**
> -   Is a transaction fraudulent?
> -   Is a prospect receptive for an offer?
> -   Which employee is assigned to handle an e-mail?

Next, breakup the question in elements needed to make the decision, also known as decision modeling. A prediction or pattern may then be used to make a decision. A machine learning task typically automates that process.

EXAMPLE DECISION MODEL USING DECISION MODELLING NOTATION

Used with permission from Decision Management Solutions.

To decide which employee is assigned to handle an e-mail depends on when the email was sent, who sent the email, what the email is about and how urgent it was. There are different kinds of elements in this list. Some of these elements are data (the email content and subject) or information derived from data (the topic of the email).

Others are rules (the policy about priorities) and some of them may be typically vague or uncertain knowledge that you want an AI model to help you with, such as detecting the topic of an email and assessing the likelihood that the tone of an email make it more urgent.

Then, analyse the potential influencers of the decision elements and phrase them as a testable hypothesis. Test them using basic statistical analysis on the available data and create a root-cause analysis. This opens the conversation and discussion in the organization and will lead to better understood and defined concepts.

If you arrived at this point it is a very good idea to revisit your concept model (step 1) while you work on the scope and task definition (step 2). Do this with your business stakeholders because they are likely to brainstorm the best ideas to analyze and test.

Finally make sure you understand the end result. Are multiple outcomes acceptable or do you need just one outcome?
Is the result of the task a classification, a ranking or a number?
Agreeing on these questions is important for your technology choice (step 4).

5.3 Collect the right data and improve its quality

The good news about this decennium is that most organizations are already collecting data. Are we collecting the right data for our objectives? Most of the time data is collected for administrative purposes and not for predictive purposes. For predictions we may need to collect more data, get new data sources or improve data quality.

If you were to ask an AI specialist today how they spend their time, the answer is likely to be 'data improvement' and 'feature extraction'.

A model to predict the orientation (North, South or East West) of a highway in the US based on the Highway number will be very complex.... while a model that is allowed to use the 'feature' indicating if the highway number is an even number (or not) will be very simple.

Note: if you do not understand this example look up a map of the US highways and try to find a pattern.

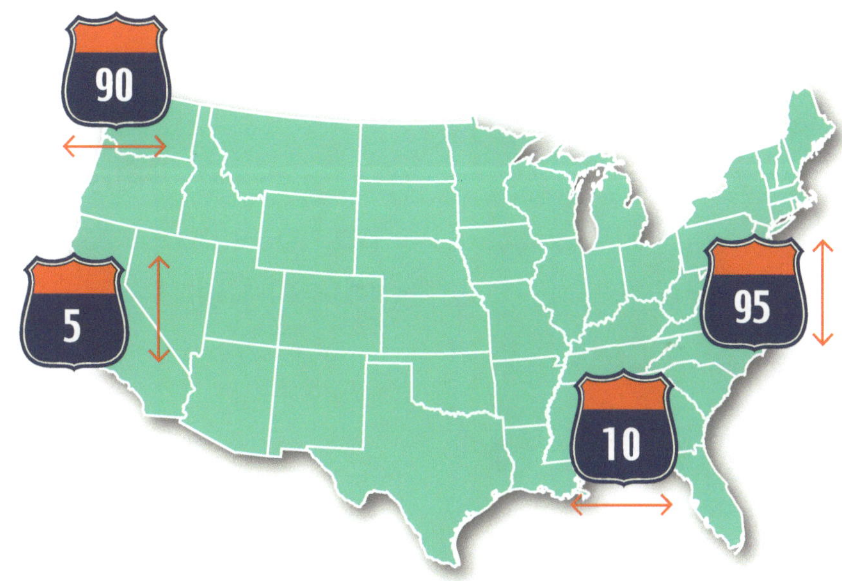

Good data quality is the most important critical success factor for AI model performance, and it is often "an issue". If the raw data does not contain the right attributes to extract the features needed to make predictions, we can enhance the raw data using rules or other AI techniques. In essence, you will need to spend time thinking about the best way to help your model by improving the data. This process is also known as feature extraction and it may be a result of your decision model (step 2). But be careful: you could introduce biases in the data.

Other AI techniques you can use to understand the data quality is to search for patterns in the data that are abnormal (anomaly detection), to look for correlations and to create clusters (features that are likely to be used together).

Collecting the right data and getting the data right may be time consuming, expensive and crucial. Therefore, you should log everything you do here (for traceability) and consider automating the workflow for next time.

"Before embarking on any ambitious project, try to kill it."
- Edsgar Dijkstra.

5.4 Select AI techniques that deliver results

Maybe you already stopped after one of the previous steps. AI is not a solution for every problem. Perhaps the problem is better solved by humans, can be easily supported with a simple set of rules or would require data that is expensive or simply not available.

To choose the right AI technique steps 1 to 3 are crucial since they help us determine what kind of data we have as input for our decision support task, how the data elements relate, what kind of decision-outcome we accept and what we already know about the task that needs to be supported. There are many AI techniques to choose from, and which technique you apply depends on the characteristics of the task and data.

It is the task that determines which technique should be used. Use a hammer if you are dealing with a nail and don't try the screwdriver because it is available! The best-known technique may not be your best choice!

Before going deeper into AI technology, let's answer the question what AI is. I studied AI when browsing was called surfing, computers used floppies and I had to explain what it was all the time. What do you think AI is?

For my mother, when asked what I do, AI is still "something with computers" and her friends are likely to be satisfied with that answer.

For most people not familiar with the topic, AI is about mimicking human intelligence with a computer such that it's 'doing something really smart', like playing chess or forecasting – things that people generally believe to be difficult.

For an analyst, AI references a family of algorithms that use a lot of data, computer processing power to find patterns in the data.

For a scientist, AI references a research area of using computers to mimic human intelligence; it is related to understanding human intelligence but does not need to use the same (biological) methods.

Before I introduce you to the technologies that are popular today in AI, I will give you a broader understanding of where these technologies come from: AI research.

AI research became a discipline after the Second World War. The topics that have been researched changed over time and involved different disciplines, such as:
- Logic and mathematics to create decision support systems
- Philosophy to understand the nature of conscience, ethics and human beings
- Psychology to improve our understanding of how cognitive tasks may be performed (both by humans and machines)
- Sociology to understand human-machine interactions
- Language theory to understand how semantics can be captured and evolved

Typical examples of intelligent tasks that have been automated using AI technology are:
- **searching for similar documents or pictures**
- **finding an optimal schedule for a school**
- **finding the shortest route between multiple places (travelling salesman problem)**
- **forecasting the weather**
- **playing chess against a computer**

The research, so far, has resulted in technology that we use daily. Many of us are not aware that we are using AI technology when searching the internet, using our credit card, or chatting with a chatbot. AI technology is all around us.

The irony is that once AI mimics an intelligent task successfully, we often do not find it intelligent any more: the target of AI is a moving horizon. Remember how frustrated Kasparov was when he was defeated by the Deep Blue algorithm of IBM? Later Google showed that computers could even play GO. What is next?

Since we relate the definition of AI to human intelligence, and there is no finite list of tasks that are considered to be intelligent, AI research will never be complete or finished. We will instead continue to find better ways to use machines to do complex tasks that are otherwise considered intelligent when performed by humans. Computer programs may reach impressive performance on some of these tasks and even outperform humans in some ways.

We need IT to do AI:

Database technology and SQL have a basis in AI research; however, I don't consider a database or any system that collects, stores and retrieves information as AI or a decision support system. At the same time: we can't do machine learning without the systems to manage the relevant information.

What does AI bring to IT?

A workflow system that ensures all enterprise assets are registered (ERP) from beginning to end, supporting multiple divisions to collaborate uses AI technology, such as rules and predictive models, to calculate prices and define valid status updates, turning a normal ERP application into a smart system that has the ability to optimize system behavior to reach and balance multiple goals.

Many 'intelligent' tasks are relatively trivial and static once they are analyzed, designed and modelled. To automate these tasks in the most efficient, effective and fastest way, many computers, devices and sensors should be connected. The data they collect has to be transformed into relevant and useful information. All this work is the responsibility of IT and is not AI, therefore IT is an important enabler for AI.

The technology delivered by AI research is typically classified as one of two dimensions: deterministic AI and non-deterministic AI.

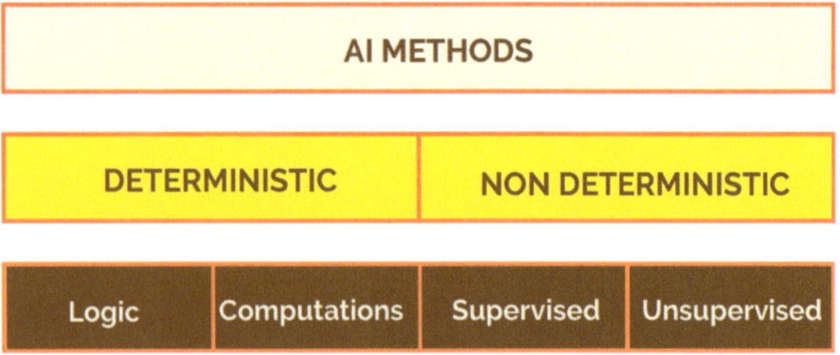

Deterministic AI methods generate results for a task using a recipe that is completely designed in advance. Given the same state or input, the method will always return the same result. The method typically uses explicit knowledge represented in

There are many problems that are non-deterministic by nature but can be well-supported using deterministic methods. Some examples include:

- Fraud detection. Although detecting fraud is a task that is uncertain by nature (we do not know what methods the fraudulent party is using in advance), most fraud detection systems do use rules to detect potential fraudulent cases.

- Traffic management. Although it is highly uncertain how road users react to traffic congestion and how congestion problems evolve, traffic managers do use rules and scenarios to minimize the consequences of traffic congestion for road users.

And luckily for deterministic AI there are also tasks that are deterministic 'by design'.
For example:

- Tax deduction calculations. The tax code uses rules that ensure taxpayers are all treated in exactly the same way.

- Claim handling. Insurance companies use rule-based software to determine if an insurance claim is eligible based on the conditions stated in the policy.

a model based on logic or computations. A rule-based system is the most well-known example. An important characteristic is that these systems do not produce a result for situations that they are not designed for. Another way they are often framed is being unable to deal with uncertainty. This is partly true. If we classify uncertainty in a quantifiable way (for example as a number between 1 and 10), we can make a deterministic system that takes this uncertainty into account; a Bayesian network is an example of such system.

Non-deterministic methods generate results for a task based on patterns that are found using advanced statistics and (big) data. The model is created using a generic machine learning method and the resulting input-output relationship is implicit. Typically, these methods generate multiple outcomes for the same state or input. Imagine that people may describe the same picture in several, slightly different ways. These algorithms work the same: they generate several descriptions for the same picture. A result could be incorrect. Outcomes are rated based on a measure such as accuracy, generality, and precision. Some results are rated better than other results and, depending on the problem domain, we may be satisfied with anything better than random accuracy or an accuracy level close to 100%.

Within this dimension an important distinction is made between supervised learning methods and unsupervised learning methods.

In supervised learning there is a 'teacher' that knows the right answers in advance; the teacher will tell the system when a generated outcome is 'right' or 'how right it is'. The machine learning algorithms use this information to guide the learning process in a certain direction. Sometimes, this may drive the results in the wrong direction. Large amounts of processing power and random variations contribute to these algorithms' learning capacity but we don't always know whether there is an optimal solution. The risk is that we can't assess if the algorithm found generic knowledge in the training data. Sometimes a model just memorizes trivial aspects of the training data, like word frequencies in image descriptions. We call this 'overfitting' in machine learning terminology. Overfitting may be detected by dividing the data in a set to train the model and a set to test the model. When the performance of the model drops on the test set, we assume that the model did not find generalized knowledge. Explanations are just as important to detect overfitting because, when the training data and the test set are very similar, performance drop is not the right indicator for overfitting.

OVERFITTING

The green line represents an overfitted model and the black line represents a regularized model. While the green line best follows the training data, it is too dependent on that data and it is likely to have a higher error rate on new unseen data, compared to the black line.

In unsupervised learning there is no 'teacher' and we do not know what the right results are in advance. Algorithms used for these tasks look for similarity in data sets, and, when presented with some input, will present similar results.

Let me say a little more on a couple of buzz-words, used for non-deterministic AI meth-ods, that you may have heard of because they are prevalent in the most recent AI hype.

Neural network is an algorithm that may be used for both supervised and unsupervised learning that is very successful for machine vision tasks, some natural language pro-cessing tasks and finding non-linear patterns in large amounts of data.

Deep learning adds many layers to a neural network or combines multiple neural networks (sequentially) to solve a task and boosted the success of speech recognition systems, cancer detection systems and im-age identification.

Random Forest is a supervised learning algorithm that generates multiple decision trees for random samples of training data. It is a typical example of a brute force method because it tries millions of random variations to successfully model a non-linear input – output relation, typically used for classification.

Non-deterministic AI methods can deliver very accurate results. The issue is that the knowledge in a model generated by a deep learning algorithm is implicit. This is a major disadvantage because it becomes very difficult to explain what the model has learned or how the result is calculated. Deterministic AI methods do not have this disadvantage: for each result a very precise recipe can be generated that tells exactly which rules are used, and in what order, to get to the result.

The solution is to combine the two methods: it is possible to train a model using deep learning and use the resulting dataset to generate a decision tree with similar, accuracy as the trained model. The decision tree consists of rules that may be used to provide explanations. For machine vision, the decision tree is likely to be a knowledge graph that explains which elements of the image have been recognized and how they contributed to the conclusion (interpretation) of the image.

Hence, XAI technology needs rule-based AI to deal with common knowledge and machine learning to deal with uncertainty, non-linear behavior and handle changes.

AI experts tend to claim that models that are good in predicting are not good in explaining and vice versa. This stance suggests that one has to choose between accuracy and transparency. I don't agree. To the contrary I claim that high accuracy and explainability can and should go hand-in-hand.

5.5   Generate good explanations

The major concern with decision support systems is that users do not trust the underlying model or prediction, and will not use it. It has been proven that a good explanation may increase the trust in a system. However, if the explanation is complicated, or nonsensical the trust decreases. So, generating a good explanation is a separate step and important.

Generating the explanation involves a second model. The explanation model has to receive information from the predictive model, get the distinguishing features, select the features that contribute to the user's confidence and present those in the right way.

There are many valuable bodies of research in sociology, psychology and cognitive science on how people define, generate, select, evaluate and present explanations. The research on explanations to date suggests that people have certain biases and social expectations about the explanation process itself.

A good explanation is built up of several components that may be perceived differently per task, domain or even the domain expert's personal preference. Practical research findings relevant to generating explanations of AI models are that explanations:

1.   Must be contrastive; by showing how the outcome is differentiated from potential other outcomes.
2.   May be incomplete; they can show one or two specific, selected causes, and not necessarily be the complete cause of an event.
3.   Must be relevant; the most likely explanation is not always the best explanation for a person and statistical generalizations based on probabilities are perceived as unsatisfying.
4.   Should be believable; they are part of a social process where knowledge is transferred in a conversation or interaction, and presented in relation to the explainer's beliefs about the explainee's beliefs.

AI experiences in healthcare to improve efficiency and effectiveness: look promosing: there is a lot of data, much uncertainty, and high costs, so every improvement counts. However, accurate AI systems are not necessarily sufficient to be used by clinical staff. The adoption has been more challenging than anticipated and the reason is that even reliable <u>machine learning</u> systems cannot earn a clinician's trust for acceptability. Research about the reasons for the lack of trust indicate that accuracy is not enough.

Clinicians know that the complexity of clinical medicine is such that no model is likely to achieve perfect predictions. Therefore, clinicians expect a non-accurate model and a system based on a model that acknowledges this challenge promotes trustworthiness.

This research on clinicians can be generalized to define what makes up a good explanation for AI models. Clinicians find it more important that the system help them to justify their clinical decision making by:

- Aligning the relevant model features with evidence-based medical practice. This can be achieved by only showing a relevant subset of the features that derive the model outcome.
- Providing the context in which the model operates. This can be achieved by showing what data is used to train the model and how certain one is that the data is relevant, complete and correct.
- Promoting the awareness on situations where the model may fall short. This can be achieved by showing certainty scores for certain outputs and indicating which combinations have not been in the training set.
- Showing a transparent design that resembles the analytical process a clinician would follow.

A good explanation uses believable model features, shows that the training data is relevant for the given situation and distinguishes the result from other outcomes. The explanation should be simple and transparent. Inspired by the <u>trust-equation</u>, I invented an equation for a good explanation:

Good explanation = (Believable + Relevant + Distinguishable) / Complexity

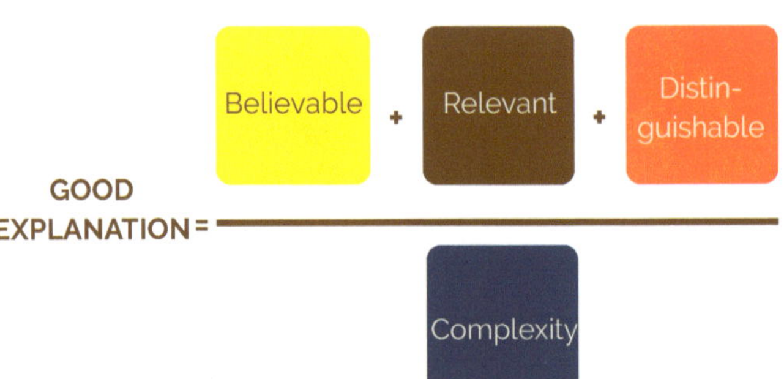

These four elements of a good explanation all converge around a single point: explanations are contextual. As a result, XAI will need a separate layer of models to generate explanations for a specific context.

Unfortunately, the AI and machine learning research community use techniques to measure overall model performance, such as accuracy and precision, as a way to explain the model. These measures will not convince a domain expert...

Instead, we need to develop separate models – also referred to as interfaces - to support the communication of an explanation. This explanation helps the user

to justify the decision. Don't forget to take into account the four elements that make up a good explanation:
-	be aligned with the user's beliefs,
-	indicate that relevant data is used to train the model,
-	indicate how to distinguish situations where the model is not applicable,
-	be transparent by being simple, following common reasoning patterns.

Nowadays, the more practically-oriented research of DARPA, the Defense Advanced Research Projects Agency in the United States, is following this approach by creating an 'explanation interface' that helps a user:
-	understand why the model produced the result,
-	understand why the model did not produce a (different) result,
-	know when the model is successful,
-	know when the model failed,
-	trust the system.

Creating a second model to generate explanations is the approach I took in the research for my graduate thesis. I first worked on a model that helped a breeder to select crops in the fifth year of a breeding process. The model was generated by a <u>genetic algorithm</u> using data from the previous 10 years. The breeder did not trust the model and sometimes had additional knowledge that was not captured in the data set; for example, that parts of the field had more shadow or mouse holes. By generating a second model, a decision tree, from the first model, using an algorithm (C4.5), we could explain to the breeder which data elements were most relevant for a specific outcome and how these data elements had been combined. That way, the breeder could understand the reasoning of the model and make his own decision, to follow or divert from the model.

The research on explainable learning algorithms can be divided in three research areas:

1. Change <u>deep learning</u> methods to ensure that only 'features' that are easy to explain are learned. Explainability, next to accuracy and precision, is one of the optimizations criteria in the model.

2. Improve techniques that learn explainable models such as Bayesian networks. An example of a Bayesian network is a well-structured model that shows symptoms and causes with their dependencies and the likelihood of the dependency in a directed graph.These causal models are easier to understand and may be used directly as an explanation.

3. Understand how to use techniques to generate explanations, such as decision tree generation, from black box models. The generated model is used to explain the result of the black box model.

Especially in domains where the costs of false positives are high (for example, when a model incorrectly classifies an applicant as being eligible or a patient as needing treatment), domain experts will be inherently skeptical about AI models. In such domains, explanations should help the domain expert to find the false positives of the AI system before these costly errors are made.

To conclude, an explanation interface should help the expert find outliers in the training data, new trends in the actual data, false positives and ideally act as a mirror, revealing the model's and the expert's decision biases. These biases, false positives, outliers and trends are the fuel for a feedback loop to evolve the system.

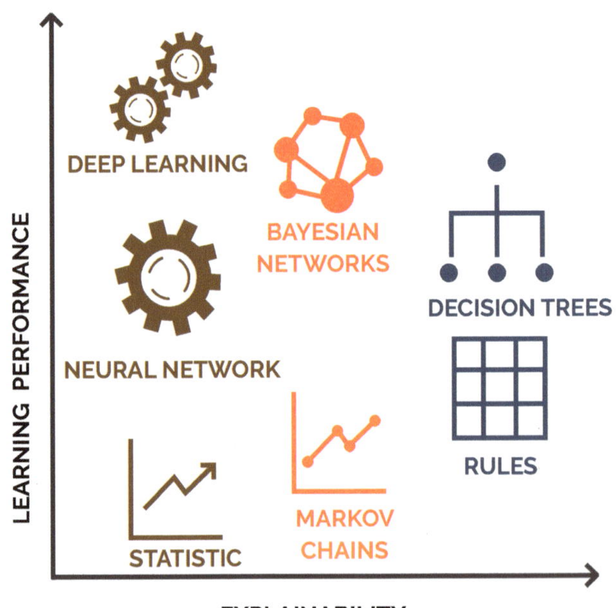

5.6   Evolve the system over time

Although a game like chess is based on relatively simple and easy to understand rules, winning at chess is not just about 'following the rules'. It takes years of experience and lots of talent to become a good chess player because there are so many different possible alternatives. By practicing you learn to recognize successful patterns. A similar reasoning may be applied for some other decision-making tasks that benefit from AI methods. Such methods explore many possible alternatives, evaluate the result using big-data, store successful patterns and are able to generate more and more reliable decision outcomes.

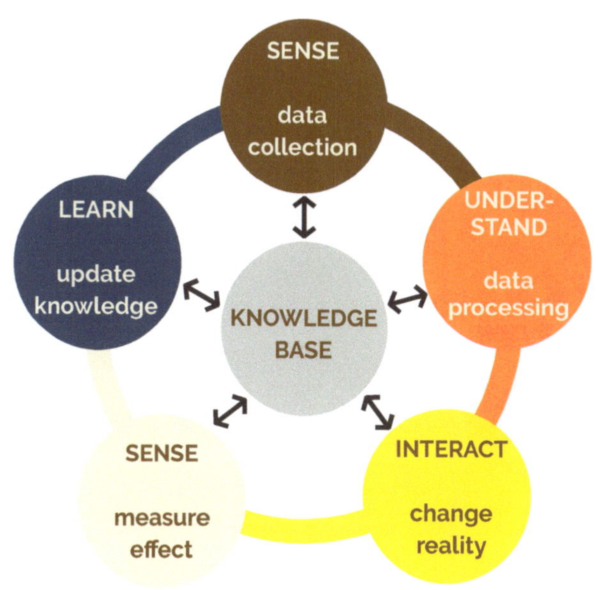

I am confident to say that we are talking about AI when knowledge is represented in such a way that it drives a process of data collection, data processing and informed decision making (interacting) to learn new knowledge.

We assume that big-data includes knowledge about the correct input-output relationship. However, what this knowledge is about remains in a black box to us. The machine-made decision maker does not reason or make inferences. When humans make a decision without inference, we sometimes call it intuition or spirituality. Do we want machines to become spiritual? In XAI, we still use AI techniques, but add an extra diagnostic feedback mechanism.

In a feedback loop, explanations play an important role. They connect understanding, sensing and learning. This feedback loop is an essential element in integrating AI in the value creation cycle of an organization. The feedback loop does not need to be fully automated. Humans need to be in the loop to validate and take responsibility. They need explanations to do so.

The idea that such a feedback loop is important for system evolution is not unique to AI solutions. It also applies to rule-based decision support systems. While rule-based systems are very common – all fortune 1000 companies have them – not many businesses are able to integrate

them well in their strategy life cycle. I have encountered businesses that don't even know what rules are included in the systems they use daily. These systems are not part of the essential value creating cycle that drives a business and I do not consider them AI.

To keep a system healthy there needs to be a way to track and analyze errors and new trends to evolve the system. Besides tracking the accuracy of the predictions automatically we need to actively look for side effects, biases and new knowledge. That is, WE, being creative and experienced business people, must be supported by a good automated workflow process to improve the decision-making process.

XAI is AI applied to automate a specific task in such a way that it is fully integrated in the business using a feedback loop with the environment.

To summarize, good XAI solutions must meet all of the following requirements:
1. Support a well-defined decision-making task, not a general task.
2. Provide an explanation and are not a black box for the end-user.
3. Use rule-based AI to include commonsense knowledge.
4. Use machine learning AI to learn patterns in data that are continuously updated when new data becomes available.
5. Monitor the performance on an operational, tactical and strategic level to exclude decision biases.
6. Are integrated in a plan do check act cycle and easy to change.

The resulting picture is an ecosystem that combines multiple levels of feedback to support the full decision-making process. It should monitor all the norms at the appropriate level and guide a learning process.

In our returning customer case study, each customer that returns but was not 'flagged' by the system as a 'returner' should be analyzed. Do we know something that the system did not know? For example, the customer returns more than 3 months later but there is a pattern. He returns every spring and only buys something for the garden…. if it makes sense, such a pattern should result in an update of the system's knowledge or system's scope description.

6

Society is demanding transparency

Artificial Intelligence technology can be used in many ways. It can fake images, fake videos and can make fake news. On the other hand, it can detect lies, detect cancer and guide people using large amounts of data to make better decisions.

AI technology can be used for the good and for the bad.

This can't be a surprise because this holds true for many other technologies and tools that humans have invented. A simple tool such as a knife may be used to cut paper, eat a meal or kill a person. A technology such as radioactive radiation may be used to cure cancer, create nuclear power or make a nuclear weapon for mass destruction. Depending on the circumstances we may judge the situation to be good or bad.

"The increasing importance of AI for the financial sector also invites us to rethink our traditional supervisory paradigms. When it comes to AI, strong operational controls could be more important than capital and liquidity buffers."

– The Dutch Financial Authority (DNB), report on the use of Artificial Intelligence in the Financial Sector, 2019.

As a society we have developed a framework of moral principles to guide our behavior and help us make judgements about good and bad. The resulting institutions and methods are used to protect humanity from harm. Recent developments in AI have raised concerns over the ethical aspects of some technology. Although AI is just 'software' and has to comply with existing regulations for software technology, even technology visionaries, such as Elon Musk, Bill Gates, Steve Wozniak and Stefan Hawking are warning that super-intelligence may be more dangerous than nuclear weapons.

Our existing framework of ethics to protect humans against harm may not be good enough. Therefore, we are developing additional guidelines, potentially resulting in new laws, to protect our society for the potential undesired effects of AI technology.

Businesses should be prepared for a society demanding more transparency. This journey starts by having a deep understanding why people and governments are concerned.

I also recommend the use of the 6-step method proposed in the previous chapter:

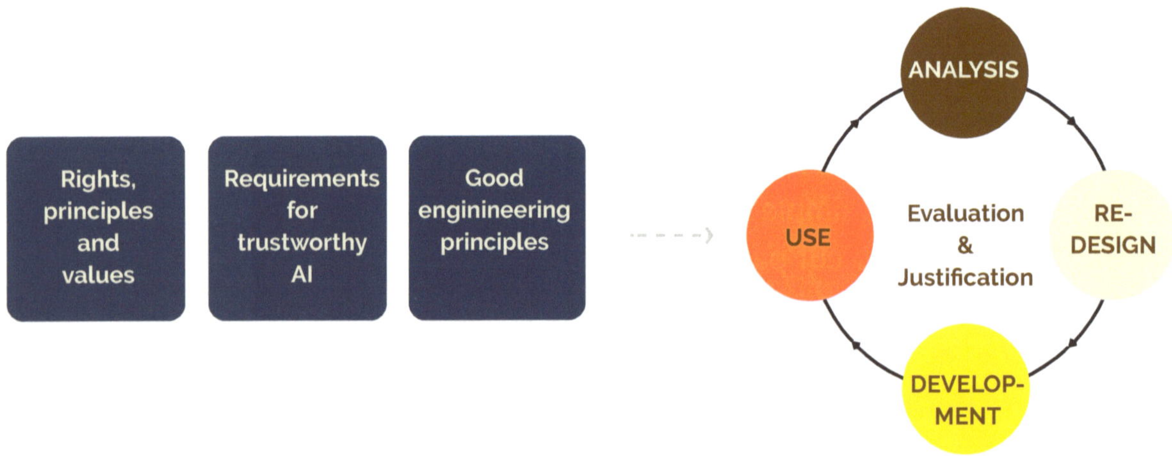

Inspired by Figure 3, realizing trustworthy AI throughout the system's entire life cycle, in the Ethics Guidelines for trustworthy AI, EU publication 2019

it will help your business to be prepared or even be a front-runner accommodating these changes. Furthermore, I recommend that you continue – or start, if you did not already ;-) – using existing best practices from software engineering such as stakeholder engagement, short iterations and introduce a feedback loop in the development life cycle.

After discussing the concerns that people and governments have in this final chapter, I am ending this story with my most important do's and don'ts for working with AI in an ethical way. Your business or industry should consider including them in their code of conduct.

6.1   People have ethical concerns

Amongst the concerns are the extent to which our legal institutions are prepared to protect us. Are the methods to mitigate the threats posed by AI technology sufficiently robust? There are political concerns as well, including the job market, skilled labor and shifts in the balance of power in international relations. Some workers fear losing their job while startups do not find the right resources and compete for expertise.

We have seen similar concerns with the introduction of other technology. Well described examples are the introduction of the letterpress, steam machine, washing machines and, more recently, mobile phones. History has shown that despite the concerns, change has made the world a better place. Should we be concerned at all? Let's start by analyzing why we are so concerned.

According to the philosopher Daniel Dennet, people assign intentionality to systems whose internal workings they do not understand. Intentionality is revealed by the language we use to reference a computer. Expressions like "the computer says no", "the computer is not cooperative today", "he must be cleaned," and "my computer loves to crash" are indications that people attribute intentions to this technology. We do this to explain their behavior to ourselves in a meaningful way. Something similar happens between humans. You may not understand your best friend fully, but you do build a meaningful relationship by assuming that your friend has similar intentions as you have. So, people assign capabilities to machines – that they actually not have, feeding horror scenarios, fears and nightmares such as 'computers taking over the world' and 'computers making unwanted decisions for us'.

AI solutions that are designed explicitly to appear intentional strengthen this phenomenon. These systems exhibit complex behaviors, usually seen as the territory of intentional humans such as planning, learning, creating and carrying on conversations. They seem to display beliefs, desires and other mental states of their own.

The fact that we assign intentions to computers may explain why 1/3 of the population would like a computer saying: "I love you", as revealed by a recent public opinion study on AI. It may also explain why we feel an urge to protect ourselves from being guided, fooled, restricted or manipulated by this same computer. Hopefully the group that wants a machine to say: "I love you" coincides with the 2/5th of the population that believe "AI can behave morally" and the 1/4th of the population that believes "AI will take over the world". Why would we even consider such idea? Aren't we good enough decision makers?

No, we are not. Psychological research provides proof that human decision makers have many systematic decision biases. "Our reasoning is biased towards what we already believe", is a good summary of the research results, known under the name "behaviorism". Other more recent research explains that these biases have evolved as strategy in order to distribute the work of 'good reasoning'.

Mercier and Spencer show that our reasoning capability improves when we reason together. Each individual finds arguments for his side and evaluates arguments for the other side. They conclude that reasoning is a social process, and the lonely thinker looking for truth and knowledge at the top of a high mountain is a romantic but false idea.

To summarize, we tend to assign intentions to computers while they are not moral beings, we are not very good decision makers, and our reasoning capabilities work best in a social process. Given these observations it is a natural development to use technology to help us make better, more objective decisions.

AI technology is a good candidate to support our decision-making process. However, there are also reasons for concern when AI is applied on a large scale. Three concerns even undermine values that are fundamental for our democracy. Systems that don't explain themselves, and are not open to feedback, prevent us from learning. Since the ability to learn is an important aspect in developing and evolving

Cambridge Analytica used Facebook posts to profile a user by classifying him or her in one of five personality traits. Combined with new posts of a user, the personality profile helped an algorithm to recommend other posts or advertisements to the user that influenced their voting behavior. The key of the "success", and trigger for ethical debates, is that users were not aware that they were being profiled, how they are being profiled and that the information that was included in their feed was limited or based this profile.

our society, this is a major concern.
AI systems that learn from potentially biased historical data are biased themselves and prevent us from identifying our biases and being egalitarian. This violates a basic right in our society: that people are treated equally.

Individualized advertisements, suggested readings and homogenized shared opinions provide us with a tunneled view based on our own opinions. This limits our freedom to choose, also a fundamental right of our society.

6.2 Governments prepare legislation demanding explanations.

The European Commission takes these concerns seriously and has set the example when it comes to ethics guidelines. They recognize that "to gain trust, which is necessary for societies to accept and use AI, the technology should be predictable, responsible, verifiable, respect fundamental rights and follow ethical rules". They also warn for the "echo chamber" where people only receive information which is consistent with their positions, or reinforces discrimination, due to biased training data.

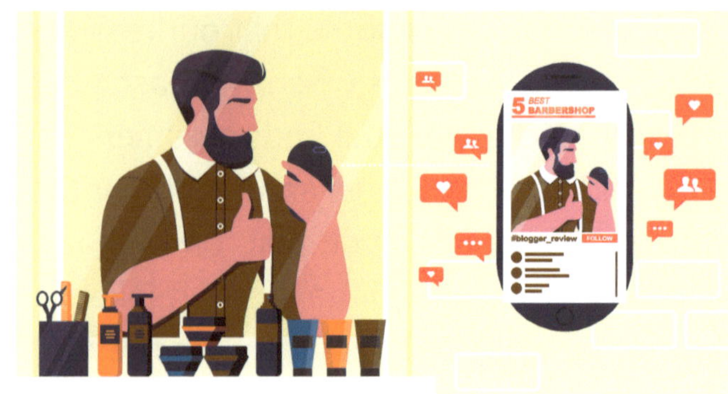

Echo chamber

The resulting ethical guidelines for trust-worthy AI, published in the spring of 2019, are organized at three levels of abstraction:

- General Terms - Ethical principles such as AI algorithms should respect human autonomy, prevent harm, be explicable and fair.
- Development Requirements - Seven requirements prescribing the way in which AI solutions are developed or on their resulting behavior. Two of these requirements relate directly to XAI (transparency and fairness)
- Trustworthiness Assessment – A list of questions to establish the trustworthiness of an AI solution.

All users and technology providers of AI solutions should be aware of these developments. Be prepared as these guidelines may become the norm mandated by certification, legislation or other legal instruments. These norms will require AI solutions to:

- trace and document a dataset and an AI system's decision;
- explain the technical processes of the AI system to humans;
- provide a suitable explanation of the AI system's decision-making process when an AI system's decision has a significant impact on people's lives;
- avoid biases in the data set, intentional exploitation of human biases and unfair competition;
- be implemented as a result of a user centric design process which engages stakeholders throughout the system's lifecycle.

The finance sector and pharmaceutical industries have already implemented some of these requirements in their code of conduct. For example, a loan assessment must be explainable. Also, the automotive industry is on its way with new rules for

self-driving cars. The code of conduct for the digital industry is still not available. It is one measure that many researchers, politicians and leaders in the industry have asked for. Anyone working with sensitive data and algorithms that guide behavior should follow such a code. Similar to the way we authorize professionals in project management, medicine, accounting and law.

THREE DOMAINS OF SOCIAL COHESION AND THEIR RESPECTIVE DIMENSIONS

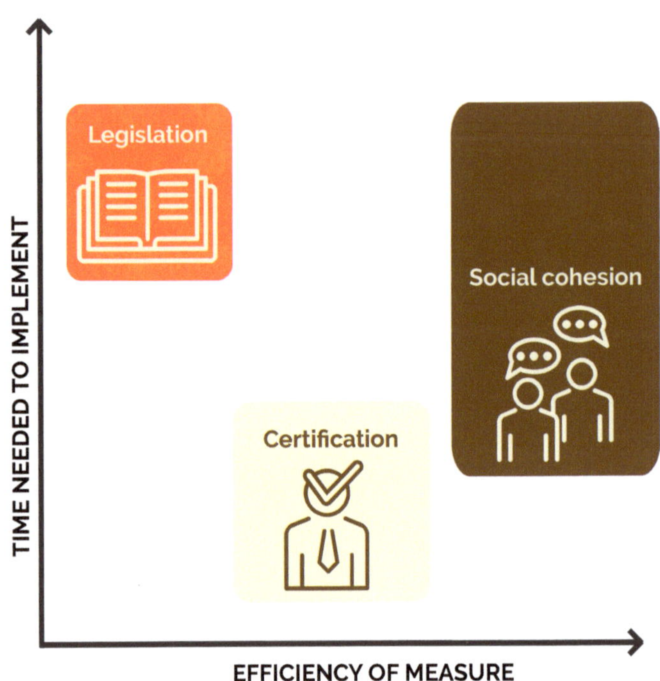

Governments need to deal and balance ethical concerns at different levels. Recently I met an ethical hacker whose work confronts him all the time with the practical reality that business and governments have commercial and political interests, and he has concluded that ethical interests are at odds with commercial and political interests. Governments are investing in digital welfare and commercial organizations only invest in security to protect privacy when the risk of high fines threaten the returns of their business.

The measures that Governments want to implement differ in effectiveness and speed of implementation.

The question is: "Do we have enough time to protect us from an Orwellian society?".

We concluded that social cohesion is potentially the strongest measure due to its distributed nature, it is efficient and difficult to remove once it implemented – which does take some time. Ironically, social cohesion is low in western countries and difficult to develop because it requires individuals to have knowledge, be connected and trust their environment.

Since such a code is not yet in place - and it will take time to implement it with sufficient support from society and professionals – how is your digital project going to be ethical? Google the internet and you will find many suggestions for elements that should be part of such code. Based on these I present my top 3 do's and don'ts to decrease the concerns related to using AI technology. But don't forget to continue, or start, following good engineering principles.

Top 3 Do's:

1. Create a feedback loop between the human and the machine requiring an explanation for each decision that guides human behavior.
2. Balance different decision criteria and anticipate changes using a modular and decentralized architecture for decision support systems.
3. Educate people about algorithms, trustful AI and critical thinking skills.

Top 3 Don'ts:

1. Don't use data that is collected for purposes other than machine learning, without making a decision model.
2. Don't filter information without making a user aware of the filter and providing the ability to remove the filter.
3. Don't hide the data that has been used to feed, train and test algorithms.

Needless to say? Don't trust an oracle.

7
Terminology

AI

The common acronym for artificial intelligence.

AI research

Any research to automate or optimize tasks that traditionally have required human intelligence.

AI technology

Any described algorithms, eventually wrapped in as a software product or service, that is the result of AI research.

Bias

The act of, or result of, showing favoritism towards something or someone that is unfair, not rational or not based on the agreed norms.

Business Rule

A guide for behavior or an action used to communicate business knowledge.

Code of conduct

A set of rules outlining the social norms, responsibilities, and proper practices of an individual belonging to a certain profession or working in a certain domain.

DARPA

The Defense Advanced Research Projects Agency is an advanced-technology branch of the U.S. Department of Defense.

Data

A set of values of subjects with respect to qualitative or quantitative variables. Data and information are often incorrectly used interchangeably; data becomes information when it is viewed in context or in post-analysis.

Decision

The result(s) or outcome(s) of a decision-making process.

Decision Modelling

The use of mathematical or scientific methods to determine the factors that influence a decision in order improve or optimize the performance of a process.

Deterministic

A system in which no randomness is involved in the development of future states of the system, a deterministic model will always produce the same output from a given starting condition or initial state.

Ethics

An area of study that deals with ideas about what is good and bad behavior.

Factor

A circumstance, fact, or influence that contributes to a result.

Feature

A distinctive attribute or aspect of something, typically part of a data set. Common synonyms are attribute, variable, field, characteristic.

Information

Facts provided or learned about something or someone, see also data.

Linear

The property of a mathematical relationship or function which means that it can be graphically represented as a straight line.

Model

A representation of something in words or numbers that can be used to tell what is likely to happen if particular facts are considered as true.

Neural Network

A set of algorithms or a type of computer program that is designed after the way the human brain operates to learn desired behavior by training – similar to Pavlov learning.

Norm

A statement or criterium that standardizes proper or acceptable behavior for a given situation or context.

OMG

The Object Management Group is an international, open membership, not-for-profit technology standards consortium, founded in 1989.

Root cause analysis

A systematic process for identifying "root causes" of problems or events and an approach for responding.

SBVR

The Semantics for Business Vocabulary and Rules is an OMG standard.

SME

A Subject Matter Expert is a person in an organization with knowledge of a specific task or domain.

System lifecycle

A continuous process that includes analysis, design and implementation of a system, with these steps repeated, on a when needed basis, based on user feedback, until it is decided to discontinue the system and the system's lifecycle ends.

Symbolic reasoning

A way to derive information by manipulating symbols and expressions according to logical rules.

Task

An activity that someone or something (when automated) performs, typically using some information and producing new information or a result, in this book we use the word task often for the task of 'making a decision about something'.

Terminology

The words and their meanings related to a particular field or domain, typically used by SME's.

XAI

The common acronym for explainable artificial intelligence

8
Recommendations for further reading and sources used

The arguments presented in this book for XAI are mostly based on, and inspired by, more than 20 years' experience implementing decision support systems in many different kinds of organizations and different domains. Another important source of inspiration came from many discussions and interactions with peers at work and conferences. Of course, I also used information from internet, articles and books. The amount of online information about AI and related topics is huge.

In this final chapter I present material that I used organized per topic. Some of that material is very well suited if you would like to explore a topic in more detail or want to understand the underlying research behind some of my claims.

Biases in data and decision making

It is very likely that you have been both the victim and cause of biased behavior. The most cited research in this field is by Kahneman [Thinking Fast and Slow, 2012] who, from a psychological perspective, questions the prevailing assumption in economics that people operate as a rational agent. Another form of prejudice is the way in which we blindly trust numbers without questioning their objectivity and understanding the story behind the number. Sanne Blauw [The number bias: how numbers lead and deceive us, 2019]

advises us to think more wisely. This all sounds very alarming. Mercier and Spencer, evolutionary biologists, therefore wonder how this human function, reason, has developed when it performs so poorly. In their book [The Enigma of Reason, 2017] they argue that reason developed not to enable us to solve abstract, logical problems, but to resolve the problems posed by living in collaborative groups.

For an overview of biases in an operational setting, like the example of gate planning in an airport, I used the article of F. Gino and G. Pisano, „Toward a theory of behavioral operations," [Manufacturing & service operations management, pp. 676-691, 10 2008], which is easy to find online. Ideas to reduce these biases in strategic decision making are presented in the McKinsey Quarterly report [The case for behavioral strategy, Dan Lovallo and Olivier Sibony, 2010] and have inspired me to introduce the idea of normware in chapter 2.1 (explanations make a solution easier adaptable to change) and 4.1 (unwanted decision biases come to the surface).

The human biases of past decision making and behavior are reflected in historic data. When this data is used by machine learning algorithms the resulting models reflect these biases. For example, visual recognition systems using pictures that contain people, risk amplifying societal stereotypes by over associating protected attributes such as gender, race or age with the target prediction. So, in a picture showing a person and cooking attributes, the model predicts that the person is female, even if the person is male. Research shows that it takes a lot of work and creativity to remove these biases. Even when you 'balance' a dataset – such that each label co-occurs equally with each gender – the learned model will be as much biased as if data had not been balanced! When I read the research on this topic from the Allen Institute for Artificial Intelligence on Fairness , I was wondering if the work to prevent biases is justified by the added value. Definitely this research is a must-read if you know your data is biased and want to mitigate your risks.

Business rules

The second AI wave, in the eighties and early nineties, was driven by expert systems based on rules. Rules are statements that constrain thinking or mandate possible actions under certain conditions. It is a way to express business knowledge and several systems, such as databases and rule engines, have standard algorithms to process rules.
An example of a mandate is expressed by the rule: "A gold customer must be offered an upgrade". In the OMG standard Semantics for Business Vocabulary and Rules (SBVR) this is named an operative business rule. SBVR differentiates them from structural business rules that state a necessity of possibility, for example: "A customer always has an account manager".
The RuleSpeak guidelines [www.rulespeak.com, 1996] help to express business knowledge as natural language rules and the business rules manifesto, more than a decade old but still relevant, explains why business rules are an important business asset. Ronald

G. Ross is behind these initiatives and author of many books on the topic. There are countless examples of rule-based systems in all industries although they are sometimes known under different names. For example, banks have mortgage assessment systems, tax administrations have income tax calculation systems and airports have gate allocation systems. All these applications use rules.

Cambridge Analytica

The idea to make psychological profiles based on questionnaires is widespread practice in psychological research and practice. A researcher from Cambridge University used this practice and had the opportunity to combine the respondent's answers on the OCEAN -Openness, Conscientiousness, Extraversion, Agreeableness, and Neuroticism - personality test, with their Facebook profile. He developed a quiz app and asked each respondent to give the app access to their Facebook profiles and those of their friends. Over 270,000 users took the quiz, which turned out to be enough to train a model that could predict the psychological profile for a Facebook profile.

The model was sold to Cambridge Analytica who used it to create a psychological profile for 60% of the American voters on Facebook. The combination of research about the OCEAN personality traits, a creative election campaign team and analytics of click behavior resulted in manipulation of voting behavior in a way that was not transparent for the Facebook user.

What surprises me most is that the University of Cambridge has survived the storm of media attention about this event without any reputational damage. Manipulation and psychological profiling are well

researched topics. Mr. Kogan, being the researcher at the University, should have had enough training in ethics to know that in a democratic system such knowledge may not be used to target voters intentionally, to influence their voting behavior and, that democracy requires a strict separation between the media, commerce and politicians.
The fact that there was something wrong is confirmed by the fine imposed on Facebook. This fine of 5 billion USD, the highest fine ever handed down by the FTC for violation of consumers' privacy, settles complaints from users. The regulator concluded that users have been "deceived" about their ability to keep personal information private.

There are many sources about this topic and all details have, up to today, probably not been revealed. A listing of facts that I used is provided by the website Dig. Watch and, of course, always check multiple sources when looking for information on the internet.

Concept model

To understand a domain, you just need to know the important concepts and their relationships, is what one of my professors told me. Concept modelling is about doing just that. It is a representation of a system, made of the composition of concepts which are used to help people know, understand, or simulate a subject, that the model represents. It is also a set of concepts. The SBVR standard from OMG provides a good basis for the semantics of a concept model. Read Rob van Haarst for an in-depth intro [SBVR made easy, 2013]. Concept modelling has also been promoted as a basis for database modelling and software development by Nijsen and Halpin [Conceptual Schema and Relational Database Design, 1989] and Ronald G. Ross [Business Rules Concepts, 2013].

Deep Blue

IBM's chess-playing computer system, Deep Blue, was the first to win both a chess game and a chess match against a reigning world champion under regular time control. It used a well-known search algorithm (alpha-beta), its success should be attributed to using this algorithm on a parallel computer. So, massive search enabled by computer processing power defeated the world champion. This approach is not related to 'deep learning', which emerged a decade later. I recommend the short Ted Talk by Garry Kasparov, [Don't fear intelligent machines. Work with them] to understand how he has experienced his defeat.

Deep learning

A relative new term referencing neural networks in a complex layered architecture: multiple neural networks may be combined, or one neural network has more than three layers. In general, it is difficult to understand what each layer learned and how the layers interact even for specialists who train the models. The YouTube channel '3Blue1Brown' created a number of animations to show the role of layers in training a neural network.

Decision model

A decision model describes the optimization and management of the thousands of automated, operational decisions that organizations make every day. Examples include the targeted pricing of products, determining which customers get automatic approval, customizing website navigation and content to satisfy regulatory mandates.. All too often these decisions are not explicitly managed, assessed or even visible to the company's business experts. Instead they are buried in the company's software code and policy manuals, where they are hidden from view and may even be contradictory. The Decision Model Notation (DMN) is an OMG standard that has increased the interest in this process in administrative organizations. For further reading follow James Taylor, author of multiple books in this field.

Engineering practices

Principles to design and build machines, structures, and other items, including bridges, tunnels, roads, vehicles, and buildings are engineering practices. Since the discipline of engineering encompasses a broad range of specialized fields, all applying mathematics, it includes software development. The success of engineering practices in the software industry has been questioned since many software projects fail. Of course, there have been many failures in the past when making bridges.... do we just need to learn more and get better or is there a fundamental difference between building bridges (or similar artefacts) and building software? A famous article by Brooks [No Silver Bullet, 1986] discusses this topic and concludes that software development is essentially complex and no 'silver bullet' can be expected to make it easier. He then goes on to describe some best practices that help to create more reliable software

Genetic algorithm

A genetic algorithm is another machine learning method for solving both constrained and unconstrained optimization problems; it is a metaheuristic inspired by the process of natural selection as described by biologists. The genetic algorithm repeatedly modifies a population of individual solutions. The solution may be in any form, for example a formula or decision tree, depending on what is expected to work for the problem at hand. At each step, the genetic algorithm selects or creates individual solutions at random to be parents and uses them to produce new solutions, re-using the 'good' elements from the parent solutions. These children solutions are the next generation. Over successive generations, the population "evolves" toward an optimal solution. Genetic algorithms have been applied in many different problem spaces and using many different solutions. A classic in this field is from John Koza [genetic programming, 1999].

Intelligence

I have been emphasizing the importance of a good overview of the terminology and concepts of a domain for understanding and successful software development. Another good practice is to not only preach best practices, but also apply them yourself. "Eat your own dogfood" is what we say in IT. So, I should have started with a definition of 'intelligence' as I write about AI to practice what I preach. However, that could have been a book in itself. I was happy when I met Dagmar Monett, an AI Researcher from the early days, who just finished a paper named "Understanding artificially intelligent systems starts with defining them: quality criteria for definitions of intelligence". The paper not only gives insight in useful definitions of intelligence but also in the process to construct definitions.

Lovelace

The fame of Lady Lovelace seems to be on the rise. Last year, I received a graphics novel as a birthday present named "the thrilling adventures of Lovelace and Babbage". Besides a factual overview of her life events, the events give inspiration for a fantastic story in which the computerization of the world is told from different perspectives using funny graphics. [Sydney

Padua, Penquin books, 2015] Another book that helped to rise her fame is having her listed in the book "computer pioneers" [Source Wikipedia, Books LLC, 2011] and the overviews of female scientists in the Spanish book " ¿Ni tontas, ni locas?: Cuando anónimo era sinónimo de mujer" [Javier Sanz & Ballesteros Díaz, Rafael, Libros Singulares, 2018]. Being a woman myself and having experienced how biased elementary school teachers rate mathematics results differently for boys and girls, I believe these 'role modelling' books, similar to "woman in tech" [Tarah Wheeler Van Vlack, Amazon Services, 2016], are important to share.

Machine Learning

Artificial Intelligence is a research area and machine learning is one of the research topics. It is a subset and studies the use of algorithms to create a computer program that performs a specific task, without using explicit instructions, relying on patterns and inference instead. The research resulted in a widely used set of algorithms, such as neural networks, genetic algorithms and some statistics-based algorithms, that can be applied in many variations to automate different kinds of tasks.

An organization that consistently uses the word Machine Learning, while many use the word Artificial Intelligence for the same technology, is BigML. I liked playing around in their cloud-based machine learning platform using a free account. Their ambition is to make machine learning very simple and indeed, what a joy they gave me! I prepared a dataset and learned about the distribution of variable values in minutes, something that would have taken me a day when I studied. They offer all the public algorithms available today in a smart configurable way.

Machine Vision

In radiology, machine vision algorithms outsmart humans almost all the time and google uses similar technology to create a description for every image you use on the internet.
Albert Gatt from the University of Malta disclosed that a neural network, with a sophisticated layered algorithm rarely seen in commercial settings, would barely drop in performance when presented with a foil, while if the images really contributed to the task you could expect the model to drop in performance [Grounded textual entailments, Proceedings of the 27th International Conference on Computational Linguistics, 2018].

There are many other examples that show that models can be fooled and these machine vision systems have limitations: a Tesla car hitting a white truck recognized as clouds, the picture of the Analytical Engine used in chapter 2 receives a description that is too general, black people classified as gorilla's.... are just a few examples. The question on the table for the research community is whether a better understanding of how these machines work will help us make these systems more reliable.

Neural Network

To get a better understanding of how a Neural Network learns, the YouTube channel 3Blue1Brown has a nice series of animations explaining the 'hello world' version of a Neural Network: recognizing the numbers 1 to 9.

Normware

A term introduced by Tom van Engers in response to the introduction of AI-based black box models in organizations that deal with operational decisions that are based on a policy or a law is normware. He relates norms to the different levels of decision making that is very common in organizations: strategic, tactical and operational in his article based on research from the University of Amsterdam and Leibniz Institute [The role of Normware in Trustworthy and Explainable AI, 2018]

Privacy

The European General Data Protection Regulation (GDPR) requires that the re-use of personal data for another purpose than the purpose when the data was initially collected, is not allowed without asking the permission of the person involved. Many AI engineers find this difficult and say: but the thing with big data and AI is that we look for patterns, but we don't know in advance what we are looking for. A good overview is given by the UK information commissioner's office (ico.org.uk) listing "Rights related to automated decision making including profiling".

Public opinion on AI

I heard Rob Walker - a former colleague, now with Pegasystems, , in a keynote about AI at Pega World, Las Vegas, 2019 - quote a public opinion study that they performed. With his permission I have been using their numbers. The public opinion, has been performed by Savanta in April 2019 and has a representative group of 5045 respondents from USA, UK, Germany, France and Japan.

Random forest

To understand why random forest is better than a single decision tree imagine you have to predict the stock market. You have 100 analysts without prior knowledge who are allowed to learn from a dataset of news reports. You know that the reports are likely to contain noise in addition to real signals. Therefore, you provide each analyst with a random subset of the data. The analyst will make a decision tree and potentially be swayed by irrelevant information. Hence, the analysts will come up with differing predictions for the same data. As a result, each individual analyst has high variance and would come up with drastically different predictions if given a new training set of reports. The idea is to not rely on any one individual but pool the votes of all the analysts. This is similar to reality when we rely on multiple sources to decide what to do. Not only is a decision tree intuitive, but so is the idea of combining them in a random forest. This analogy used in this explanation is based on an article from Will Koehrsen [An Implementation and Explanation of the Random Forest in Python, Medium, 2018]

Research on explanations

My experiences with generating an explanation for automated decisions were confirmed when reading the interesting overview of research by T. Miller [Explanations in artificial intelligence: insights from the social sciences. August 2018].
More detailed research, such as the survey about AI for clinicians [S. Tonekaboni et al, What Clinicians Want: Contextualizing Explainable Machine Learning for Clinical End Use, 2019] is needed to advance the industry.
The need is clear, the solution is not obvious, as is concluded by lawmakers struggling with legal decision making and AI [Christopher Markoe and Simon Deakin, Ex Machina Lex: The Limits of Legal Computability, 2019].

There is a need for methods to present explanations and prevent businesses to re-invent the wheel every time. Representations that can be presented readily to a user as an explanation are decision trees, business rules and linear regression models. All representations will indicate the importance of a feature for the individual result. A representation has a complexity measure, for example by setting a maximum depth to a decision tree or a maximum number of concepts in a rule. Here are some technologies and tools to consider:

-       Lime [https://github.com/marcotcr/lime] is the name of an open source toolbox that you may want to explore if you need explanations for text classifiers or image recognition software. It can generate explanations for a black box model by creating random sample decisions and using the result to weight the relevance of features while minimizing the explanation complexity. For text classifiers a list of the most relevant words is presented in order of importance. For a picture it shows the most relevant pixels of a picture that contributed to the result. If, for example, the most relevant element of a picture to detect a wolf is 'snow'.... then you may not trust the model!
-       Decision tree generation using an algorithm such as C4.5 is a step further. At some point in time it may be integrated in a toolbox such as Lime, the researchers suggest in their paper [Explaining the Predictions of Any Classifier, KDD ,2016].
-       Partial dependency plots show the effect a feature has on the outcome of a predictive based model. However, I believe that users may need much training to interpret these plots.
-       A recent development has been the implementation of Shapley values into machine learning applications. For each variable, it randomly samples other values from the data set and calculates the change in the model score. These changes are then averaged for each variable to create a summary score, but also gives information on how important certain variables are for a specific data point.

These technologies may be helpful but there will still be a lot of engineering involved to create a good explanation model.

Social Cohesion

A notion that is mostly used in the context of urban development. I believe the concept is also important when we discuss data protection laws and ethical decision making. I started to get a better understanding of this topic after reading the publication Trying Times Rethinking [Ulrike Spohn, Bertelsmann Stiftung, 2019] which was my inspiration to define the concept in terms of knowledge, trust and connectedness.

Technology trend

The recent interest in AI has been triggered by the availability of huge amounts of data and not by major technological innovations. US analysts such as McKinsey, Gartner, Forrester and Centric Digital report about the development of the trend. See the website towardsdatascience.com for an overview and search for 'artificial intelligence what it is and how the technology can be used in a business environment'.

The Startup Way

Mature organizations have a tendency to become more conservative and less innovative. The Startup Way is a system of entrepreneurial management that claims to lead organizations of all sizes to sustainable growth and long-term impact, described by Eric Ries.

It draws on the ideas behind 'lean startup' which advocates a minimal viable product, a customer-focus and uses tests to choose between pivot or persevere in a continuous innovation method.

Trust-equation

Defined by Charles H. Green, covers the most common meanings of trust that you encounter in everyday business interactions. What's important to remember is that the meanings are almost entirely personal, not institutional. The Trust Equation has one variable in the denominator and three in the numerator. Increasing the value of the factors in the numerator – credibility, reliability and intimacy – increases the value of trust. Increasing the value of the denominator – self-orientation – decreases the value of trust.

Trustworthy AI

This term is used by the High-Level Expert Group on AI from the European commission. They presented the report "Ethics guidelines for trustworthy AI" in April 2019. According to the Guidelines, trustworthy AI should be: (1) lawful – respecting all applicable laws and regulations, (2) ethical – respecting ethical principles and values, (3) robust – both from a technical perspective while taking into account its social environment.

The Guidelines also set forth 7 key requirements that AI systems should meet in order to be deemed trustworthy. A specific assessment list aims to help verify the application of each of the key requirements.

XAI technology

The amount of research in the explainability of machine learned models is increasing. If you would like to follow what is happening, you could look into the DARPA research results, and follow the publications of the Allen Institute for Artificial Intelligence. There are also open-source initiatives such as lime, PDPbox and SHAP. See an overview on the article: opening black boxes how to leverage explainable machine learning found at the website towardsdatascience. com.